To MY NE~~~~~~

YOU CAN'T HAVE EVERYTHING, AFTER ALL, WHERE WOULD YOU PUT IT?

Best Wishes

RKL

2/7/14

KARL, THE UNIVERSE AND EVERYTHING

Also by
DR KARL KRUSZELNICKI

Curious & Curiouser
Brain Food
50 Shades of Grey Matter
Game of Knowns
House of Karls
Dr Karl's Short Back & Science
The Doctor

Dinosaurs Aren't Dead
Dr Karl's Big Book of Science Stuff
(and Nonsense)
Dr Karl's Even Bigger Book of Science
Stuff (and Nonsense)
Dr Karl's Biggest Book of Science Stuff
(and Nonsense)
Dr Karl's Big Book of Amazing Animals
Dr Karl's Little Book of Space
Dr Karl's Little Book of Dinos

KARL, THE UNIVERSE AND EVERYTHING

Almost everything you need to know about almost everything

Dr Karl Kruszelnicki

Pan Macmillan Australia

First published 2017 in Macmillan by Pan Macmillan Australia Pty Ltd
1 Market Street, Sydney, New South Wales, Australia, 2000
This Macmillan edition published 2018 by Pan Macmillan Australia Pty Ltd

Text copyright © Karl S. Kruszelnicki Pty Ltd 2017
Illustrations copyright © Jules Faber 2017

The moral right of the author to be identified as the author of this work
has been asserted.

All rights reserved. No part of this book may be reproduced or transmitted by any person or entity (including Google, Amazon or similar organisations), in any form or by any means, electronic or mechanical, including photocopying, recording, scanning or by any information storage and retrieval system, without prior permission in writing from the publisher.

Cataloguing-in-Publication entry is available
from the National Library of Australia
http://catalogue.nla.gov.au

Cover and text design by Alissa Dinallo
Typeset in Electra by Alissa Dinallo

Printed by McPherson's Printing Group

Material on page 38 from *Flyaway* by Desmond Bagley reprinted by permission of HarperCollins Publishers Ltd © Desmond Bagley 1978
Material on page 40 from "Want to Block Earworms from Conscious Awareness? B(u)y Gum!", by C. Philip Beaman et al., *The Quarterly Journal of Experimental Psychology*, copyright © The Experimental Psychology Society reprinted by permission of Taylor & Francis Ltd, http://www.tandfonline.com on behalf of The Experimental Psychology Society.
Material on page 126 reprinted (adapted) with permission from "Material Degradomics: On the Smell of Old Books", by Matija Strlič et al., *Analytical Chemistry*, January 2009, Vol. 81, No. 20. Copyright 2009 American Chemical Society.

We advise that the information contained in this book does not negate personal responsibility on the part of the reader for their own health and safety. It is recommended that individually tailored advice is sought from your healthcare or medical professional. The publishers and their respective employees, agents and authors, are not liable for injuries or damage occasioned to any person as a result of reading or following the information contained in this book.

The author and the publisher have made every effort to contact copyright holders for material used in this book. Any person or organisation that may have been overlooked should contact the publisher.

Papers used by Pan Macmillan Australia Pty Ltd are natural, recyclable products made from wood grown in well managed forests. The manufacturing processes conform to the environmental regulations of the country of origin.

Life. That's what I dedicate this book to.

It is a word that is missing from the title of this book, but I love it.

It's bursting out of the 5-metre-long root of a tree that is pushing up and wrecking the pavement in a city street. It is, and should be, chaotic and messy.

Even though it's so vibrant and forceful, it's pretty hard to define. Animals, plants and bacteria are alive, but what about viruses? And we still don't know how a bunch of inanimate chemicals (fats, proteins, carbohydrates, water and various minerals) turn into a living creature that can reproduce.

Earth is the only place where we know for sure that Life exists. Personally, I'm pretty excited to be living – and on the right side of the grass!

The Next Really Big Thing that I'm hanging out for is finding Life in the skies. Hello, Enceladus!

CONTENTS

01 SEXIST AIR-CON 1

02 COFFEE NAP 10

03 POO IN 12 SECONDS 14

04 BRAVE NEW WOMB 24

05 EARWORM 36

06 HYDROTHERMAL VENTS AND INVISIBLE MOUNTAIN RANGE 42

07 LIFE ON ENCELADUS 54

08 REFRIED BEANS 70

09 MISOPHONIA – HATING SOUND 72

10 SHOUTING AT HARD DRIVES 80

11 PONYTAIL SWING 88

12 LIGHTNING POWER 98

13 TRUCK SHIFTS DATA FASTER THAN OPTIC FIBRE 110

14 THE SMELL OF BOOKS 114

15 EMERGENCY EMERGENCY PHONE CALLS 128

16 LUSCIOUS LIPS 130

17 SOLAR ENERGY PAYBACK TIME..136

18 POKIES, TURNING THE TABLES................................ 140

19 GOLD NOBEL MEDAL INVISIBLE
TO NAZIS 154

20 SPACE JUNK...................................162

21 CHILDHOOD AMNESIA 184

22 NATURAL ALARM CLOCK.......192

23 SINKHOLES 194

24 EARTH STOPS SPINNING ..204

25 CANE TOADS CONFIRM
CONCEPTION........................... 214

26 CHRONIC LATENESS 218

27 PB/5 PEDESTRIAN BUTTON 228

28 BEER "BEATS" CANCER?........240

29 MARIJUANA FOR MEMORY &
LEARNING?..242

30 HOTEL MAGNETIC CARD ERASURE ... 250

31 IG NOBEL PRIZES 2016 254

32 WIFI SPY 264

33 SPIDER WEBS DON'T TWIST 274

REFERENCES 281

01

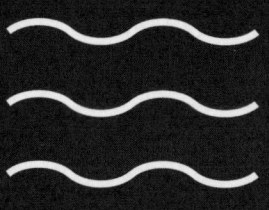

AIR-CONDITIONING MAY be "essential" in our modern world – but it comes at a price.

There is the obvious up-front purchase cost, as well as the cost of the energy used to cool our buildings. But what about the hidden Gender Imbalance Cost? You see, modern air-con set-ups all share the same fundamental flaw – they are sexist.

In the land of air-con men are fine in shirt-sleeves, but women often shiver and have to wear cardigans and scarves.

AIR-CON HISTORY

So first, let's get some context.

In the good old days, before the Biblical "Fall From Grace", the weather was always fine. In Milton's *Paradise Lost*, the most that Adam had to do to stay perfectly comfortable was to seek a little shade around midday.

But then, things got messy!

The cause was that unfortunate incident involving Adam, Eve, the apple and the talking snake. So God got mad, and tipped the Earth off its vertical axis – to punish the lot of them. This was why – according to *Paradise Lost* – Adam and Eve then suffered "scorching heat in summer . . . pinching cold in winter".

(In my 24th book, *Dis Information and Other Wikkid Myths*, I wrote that over 80 per cent of science graduates from Harvard did not know that the tilt of the Earth is the Reason for the Season.)

Humanity moved on from Adam and Eve, and started using Science to manage the summer heat.

Five thousand years ago, the Ancient Egyptians cooled down using the evaporation of water. They hung moistened reeds in windows, for the wind to blow through. This provided both cooling and a welcome increase in humidity. A few thousand years later, the

SEXIST AIR-CON

Ancient Romans circulated the water from their aqueducts through the walls of the houses of certain privileged people. (Even then, it was cool to be cool.) The Persians used wind towers and cisterns for cooling in the heat of summer.

By the second century AD, the Chinese had huge human-powered fans, 3 metres in diameter. By 747 AD, they had moved to water-powered fans blowing air over water rising from fountains.

James Harrison, the "Father of Refrigeration", invented the mechanical refrigeration process for creating ice in around 1851. As a journalist in Geelong in Victoria, he noticed that ether (used for cleaning the movable type in a printing press) would leave the type cold to the touch. He invented (and patented) a machine that could make ice from water. By 1855, his machine was making 3000 kilograms of ice each day. (See my 21st book, *Great Australian Facts and Firsts*.)

After another half-century, in 1902, Willis Carrier in New York invented the first modern electrical air-conditioning. It could control both temperature and humidity. It was originally designed to give more consistency to the paper and ink alignment in a printing plant. The actual term "air-conditioning" was first used in 1906. By 1914, air-con was being used in one lucky (and wealthy) private home, the residence of Charles Gates in Minneapolis. The first portable in-window air-con unit was invented in 1945.

Air-Con Energy

Worldwide, the sales of air-conditioning units increase by 20 per cent each year. China and India are leading the charge, due to their rising affluence and rising temperatures. (See my story on Heat Waves in my 40th book, *The Doctor*.)

Currently in the USA, 40 per cent of all electricity produced goes to commercial and residential buildings. Of this 40 per cent of the energy pie, the biggest slice goes to heating, ventilation and air-conditioning.

Engineers are exploring more energy-efficient pathways in cooling. These include replacing coolant fluids with solid materials, installing specialised membranes that can cool air by condensing water, and developing individualised micro-climate systems in buildings.

SEXIST STANDARD 55

"Standards" exist for nearly every human activity, from welding to tea-making – and yes indeedy, for how to set the air-con.

For the last half-century, the air-conditioning settings in our buildings have followed the famous Standard 55. Like a lot of stuff in the 1960s this Standard turned out to be quite sexist.

Standard 55 is a set of guidelines on how to regulate indoor temperature and humidity.

It was introduced in 1966 by the American Society of Heating, Refrigeration and Air-Conditioning Engineers, and then spread worldwide. It took account of six factors – four environmental and two human.

The four environmental factors were the outside air temperature, the amount of heat energy radiating from various surfaces (including humans), the humidity of the air and, finally, the air speed.

The first human factor was related to the human Metabolic Rate. It was called the MET (Metabolic Equivalent of Task). The energy cost of a typical human activity in an office (such as filing, typing, or operating a cash register) was defined as having a MET of 1.0.

The second human factor was clothing (called "clo").

And you guessed it, the engineers who developed the Standard chose a man as their reference point. Specifically, they chose a 40-year-old man, weighing 70 kilograms, and dressed in a full 1960s business suit complete with shirt, singlet, underwear, and covered footwear. That outfit was defined to have a "clo" of 1.0.

And almost all modern air-conditioning systems are still set up to follow this famous Standard 55.

STANDARD 55 PROBLEMS

Now that you know the background, you can see a few problems.

First, each person's individual Metabolic Rate varies enormously. There are also societal factors. For example, one study found that Japanese women preferred a work temperature of 25.2°C, while European and North American men under the same conditions preferred 22.1°C. A Finnish study found similar results. Other important factors include the person's height, age, weight, fitness, gender and, of course, the type of work. A mechanic in overalls changing a gearbox in a car will have a much higher MET than an office worker at a computer.

Women usually have both lower height and weight, and a higher percentage of body fat than men. (Body fat is an insulator, but it is also less metabolically active than muscle.) Overall, women pump out about 20 to 35 per cent less heat than men do. This means that women don't need as much cooling as men.

The second problem is also obvious. Women often wear lighter clothing, and cover up less. In an office, you usually don't see a man's bare knees.

The third problem is a little more subtle. It's related to the "glass ceiling". This is the unfair social condition where women are less likely than men to get promoted, regardless of their qualifications.

Bosses (often men) get the more attractive corner offices with lots of glass and great views. Speaking generally, women are clustered more towards the centre of the building.

In most fit-outs, the centre of the building is where the air-con vents are – because they are close to the central Services Duct or Services Riser, where the lifts, electricity, water, electronics, air-con ducts and the like are all concentrated. So women, in their lighter clothes, are directly in the path of the coldest air blasting out of the vents, on its way to the warmer corner offices.

Now there's another factor to consider. Glass usually leaks heat like crazy. It's almost always cheap, ineffective single glazing, which lets in lots of heat in summer (and leaks lots of heat in winter).

So in summer, the men in their suits in the warmer corner offices want the air-con running even colder than normal to feel comfortable.

This situation means that sometimes women actually have to turn on heaters, in the middle of summer, to stop them continually shivering inside the air-conditioned office.

Yup, women literally get left "out in the cold".

THE BIGGER PICTURE

Cardigan wars aside, there's a bigger issue with air-conditioning – the cost of running it. But addressing this cost could be both a win–win situation for energy, and help end the days of sexist air-con.

If we just set the thermostat to a slightly higher temperature, we can save energy and money.

According to Richard de Dear, Professor of Architectural and Design Science at the University of Sydney, just resetting the thermostat from 22°C to 25°C could cut a quarter off the cooling bill. That's a huge saving!

At that energy-friendly temperature, jacketless men can chill out, and women won't have to battle two wars at the same time – the air-con cold front and the glass ceiling.

Air-Con for Heating?

Surprisingly, if you get what's called a "reverse-cycle air-conditioner", you can heat as well as cool. After all, air-con usually shifts heat from the inside to the outside. In the "reverse-cycle" mode, it does the opposite – it shifts heat from the outside to the inside.

It now operates as a "heat pump". Heat pumps have their limitations – for example, they work best in mild winter climates, where the temperature ranges between 4°C and 13°C. But compared to electrical resistance heating, they can use much less electricity (one third to one quarter) to provide the same amount of heat.

02

COFFEE NAP

ZZZ

WHEN YOU'RE TIRED, caffeine can feel like a lifesaver. It can improve your performance and make you more awake. So can a nap. But what happens when you combine the two "powers" – caffeine plus a nap?

If you time them correctly, both your performance and your wakefulness are improved even further.

SLEEP DEBT

If you consistently do not get enough sleep, you build up a so-called "sleep debt". Obviously, for your long-term health and performance, you need to repay this sleep debt – and sleeping is the only way to do that.

A short nap does reduce your sleep debt – and temporarily helps overcome your sleepiness.

CAFFEINE 101

The pick-me-up effect of caffeine isn't immediate.

Caffeine takes time to pass from the stomach to the small intestine, then to the liver for processing, and finally into the generalised bloodstream of the body. It takes your bloodstream about 45 minutes to reach peak caffeine blood level.

So drink your coffee, wait roughly 45 minutes, and then you'll have about 100 milligrams of caffeine circulating in your bloodstream.

But it's not a steady state. About half of the caffeine in your bloodstream is removed every 4 to 5 hours. Your circulating caffeine drops to 50 milligrams about four hours after you drank it, and then down to 25 milligrams after another 4 to 5 hours.

The "average" dose of caffeine is about 100 milligrams per cup of coffee. It's slightly less in instant coffee, tea, energy drinks (Red Bull and V) and some over-the-counter medicines (e.g. Panadol Extra).

Most adults can have 300 to 500 milligrams of caffeine each day with no ill effects – and quite probably with good effects. (Read my story "Coffee is Now Good" in my 40th book, *The Doctor*.) More than that can produce anxiety, nervousness, irritability, restlessness, heart palpitations, chills, agitation and increased urine flow.

In children, smaller doses can produce anxiety.

SMALL 1997 STUDY ON DRIVING

A small study, on only 12 university students, tested them for sleepiness. Their performance was monitored on how well they drove on a virtual reality expressway – which was deliberately mostly dull and tedious. The testing period ran from 2 to 5 p.m. – covering the infamous afternoon lull.

The study started with a 30-minute simulated drive, followed by a 30-minute break. The volunteers were cycled through three possible options during this break.

First, the placebo condition. This group got neither caffeine nor a nap. Instead, they were given 200 millilitres of decaffeinated coffee, right at the beginning of the half-hour break.

Second, caffeine only. They were given 200 millilitres of decaffeinated coffee with 200 milligrams of added caffeine.

Third, caffeine plus nap. They were given the 200 millilitres of decaffeinated coffee with the added 200 milligrams of caffeine – and also asked to try to take a nap. The caffeine didn't interfere with the nap, as its blood levels would take a while to rise – and this was also during the afternoon lull. If the volunteers did sleep, they were woken up just before the end of the break.

They were all then sent on their simulated drive for another two hours. Various unexpected events were thrown at them – such as having to avoid another vehicle. They were also monitored for how well they stayed in their lane.

The placebo group (no caffeine, no nap) had a certain number of "incidents" – let's call this 100 per cent. Caffeine alone brought the incident rate down to 34 per cent. Caffeine plus a nap brought this down to 9 per cent.

So in the short term, a Coffee Nap will help you get home safely. But caffeine will only overcome sleepiness, and only temporarily. It will do nothing for your sleep debt.

Sleep is one of those things money can't buy.

03

POO IN 12 SECONDS

THE PUNCHY ONLINE heading got my attention, all right! How could anybody resist a story titled "All Mammals Poop In 12 Seconds..."?

Wow! A 20-gram mouse and a 7-tonne elephant each take *exactly* 12 seconds to have a bowel motion. Really?

And it's the same for humans as well? *All* humans – meat-eaters and vegetarians? Even when we get diarrhoea and constipation? And with the frequency of bowel motions ranging from a few each day to one every few days? Twelve seconds? Really?

FAKE FACTS

Despite being skeptical, I was very intrigued – so I scrolled down past the clickbait headline.

Soon I was reading (with great surprise) that a mouse poo was 300 grams. But your typical mouse weighs only 20 grams. Whoa! It's clearly impossible for a mouse to defecate 15 times its own body weight! (Where would it store the poo?)

Still distracted by the mouse-poo dilemma, I read on. Next came the claim that an elephant poo had been weighed at 3837 kilograms. That's more than half the weight of a really, really big elephant. After a morning poo that big, you'd be left with a really saggy, baggy elephant – just skin and bones.

Copromancy

We humans have studied poo for millennia.

Nearly all ancient cultures practised "copromancy" – the diagnosis of health or disease, based on the texture, shape and size of faeces.

Discussion of bowel movements is part of the modern-day Gastrointestinal Examination. In Germany, many of the toilet bowls have a flat, raised section, so you can examine your poo after a bowel motion.

So "poopular science" has been in our human cultures for ages . . .

QUICK, TAKE ME TO A LIBRARY!

Now I realised that I needed to stop reading the crappy online article. I needed to get me to the Library to find the Genuine Peer-Reviewed Scientific Paper.

It took me only a few minutes. The original paper was called "Hydrodynamics of Defecation". It was published by the Royal Society of Chemistry (a reputable source), in one of their journals called (rather coyly) *Soft Matter*. The team of authors included Fluid Dynamicists, a Colorectal Surgeon and, of course, undergraduates (also known as Cheap Labour). The undergraduates "filmed defecation, and [also] hand-picked feces from 34 mammalian species at Zoo Atlanta in order to measure their density and viscosity".

In *The Conversation*, one of the authors David Hu described how becoming a parent turned him from a "poo-analysis novice to a wizened connoisseur. [His life passed] by in a series of images: hard feces pellets like peas, to long feces like a smooth snake, to a puddle of brown water."

This real science paper was fascinating – but not tempting enough to make me want to change careers.

It reported (for those 34 mammals) that about 10 per cent of the potential energy of the food was still in the faeces. This left about 58 per cent of the energy in the food to do "useful work", while the remaining 32 per cent was used to operate the animal's body – for metabolism.

They measured that on a daily basis incoming food made up about 8 per cent of the animal's mass, but that faeces were only about 1 per cent of body mass.

More specifically, a typical elephant poo weighed about 15 kilograms, not 3837 kilograms as reported in the non-peer-reviewed literature. (That's a pretty impressively large margin of error.)

And the time taken to have a poo, or "eject a faecal piece"?

Well, about two thirds of mammals took between 5 and 19 seconds. This is not as catchy as "All Mammals Poop In 12 Seconds".

The "12-Second" headline fib is a shame. They didn't need to tell a Little Lie. They missed the point. It's impressive that most animals (covering a wildly different range of sizes and diets) could do a poo in under 20 seconds.

And that's what these scientists were interested in – why the margin for poo-time was fairly consistently narrow in mammals as varied as cats and elephants.

To try to explain this better, the scientists joined the millennia-old tradition of studying the actual poo.

Floaters and Sinkers

If we take the density of water to be 1.0, faeces with a density less than 1.0 would float – and faeces with a density greater than 1.0 would sink. The faeces of 34 mammals were measured (thank you, undergraduates), and their density was found to be between 0.2 and 1.5.

The density of faeces depends on the diet, and on how much water the animal extracts from them. For example, an animal that eats bone (with twice the density of water) would have quite dense faeces. (By the way, bone fragments have been found in fossilised dinosaur poo, called "coprolites".) An animal that lives in dry deserts would remove as much water as possible from its faeces – again making them dense.

The large carnivores (bears, tigers, lions, etc.) had faeces with an average density of 1.4 – sinkers.

On the other hand, herbivores such as pandas, elephants and kangaroos had faeces with an average density of 0.83 – floaters. These animals ate food with a very low density of nutrition, and then excreted much of this food completely undigested.

DIGESTION 101

Defecation is the final part of digestion.

Food comes into your mouth, passes through the stomach and small intestine, and then enters the colon. By this stage, the food has been turned into a wet slush, and much of its "goodness" has been extracted.

This "foody mush" gets pushed through your gut by a process called "peristalsis". Let me explain with an example. Imagine a flexible plastic pipe, about the diameter of your wrist. Fill it with jelly. Wrap your hand around it, squeeze, put your other hand next to it, and squeeze. Repeat many times, and the jelly will ooze out of the end of the flexible plastic pipe. This is the process that your gut uses to push its contents down to the rectum.

The rectum is the last section of the gut. This is where the faeces are temporarily stored. As the rectum gets fuller, stretch receptors in the walls of the rectum start firing, and eventually trigger the act of defecation.

Faeces Only in Rectum?

In medical school, we were taught that faeces were stored only in the rectum. But this study found that about half the faeces were held in the rectum, and the other half immediately above it – in the descending colon.

MUCUS IS MIGHTY

An essential part of this process is the thin layer of mucus that coats the inside wall of the colon and rectum. One of its functions is to be a wet lubricant. Check it out the next time you see a dog having a poo.

For a few moments after expulsion, the dog poo is shiny. What you're seeing is light reflecting off the mucus. Then the mucus evaporates, and the poo takes on a dry, dull appearance. The thickness of the mucus ranges from 30 microns in a small animal to about 100 microns in bigger animals. (A micron is a millionth of a metre. For comparison, a human hair is about 50 to 70 microns thick.) As the faeces moves down the gut and gets closer to the rectum, the enveloping layer of mucus gets thicker.

This mucus is the key to how it can take roughly the same amount of time for the defecation event of a mouse and an elephant.

The mucus has a strange property called "shear-thinning". It can split into layers (like a pack of cards). At the same time, these layers can slide apart from each other, while still having some stickiness (or adhesion) to each other. This strange splitting-sticking property was part of the mathematics that the fluid dynamicists had to incorporate into their peer-reviewed study. They also had to incorporate many other properties, such as the ability of the pieces of faeces to flow (or distort) under pressure.

They found that the way that a piece of faeces leaves your body is not like a thin stream of toothpaste being extruded out of a larger tube. Instead, it's more like a little bullet sliding through a hose.

Bigger animals do have bigger faeces, but they also defecate at higher speeds. Typical speeds are 6 centimetres per second for an elephant, 2 centimetres per second for a human, but just 1 centimetre per second for a dog.

Each year, gut conditions such as Inflammatory Bowel Disease and Irritable Bowel Syndrome cost billions of dollars in the USA to treat. They can cause wildly altered bowel habits, which can make life awful for the sufferers.

We need studies like this one to understand these gut conditions – and then hopefully offer effective treatments.

So we study poo to add to our knowledge of bowel basics – not just because we love mouses' faeces to pieces . . .

Slippery Dog Poo

You might have been unlucky enough, at least once in your life, to step on some dog poo.
In many cases, under pressure, it takes on the splitting-sticking behaviour of the mucus that lines the gut. And so your foot slides forward on the pavement, leaving behind a long trail of . . .

04

BRAVE NEW WOMB

I HAVE A favourite organ, and it is the uterus – because it goes through such amazing transformations. (I secretly call it the Uter-House – it's a House for the Baby.) Non-pregnant, this hollow muscular organ with a thick wall is the size of a small upside-down pear, weighing less than 60 grams. But by the end of a typical pregnancy, it's the size of a shopping bag, weighing about a kilogram – and it's packed fairly tightly with a bonny, baby human. The cavity inside has increased in size by 500 times.

The natural uterus is one of the Wonders of Nature and very hard to reproduce. *The Artificial Uterus* has been a sci-fi staple for years (think of the Hatcheries in *Brave New World*). After half a century of trying, we are getting close to making it real.

Beginning in the 1950s, people explored different ways of making an Artificial Uterus, for use in emergencies in the late stages of pregnancy – long after the organs had been formed. They eventually realised that there were four main problems to overcome.

The first problem was making an adequate placenta. This Artificial Placenta would have only two jobs – oxygenating the baby's blood, and providing a few essential nutrients.

The second problem was how to connect this Artificial Placenta to the unborn baby. This was solved by designing a biomedical direct connection device to the umbilical cord – between the Artificial Placenta and the baby.

The third problem was that previous attempts had used a pump to move blood into, and out of, the baby. Unfortunately, this pump had damaged the baby's own heart. This was solved by inventing an oxygenator with very low resistance to flow. This meant that the tiny heart of the baby could pump the blood by itself, and that there was no need for an external pump.

The fourth problem was how to deal with the amniotic fluid. This was solved by having synthetic amniotic fluid, in a semi-closed system. The baby was popped into a "BioBag" and sterile fluid was slowly infused into one side of the bag. At the same time, the older fluid flowed out the other side.

Why I Love and Admire the Uterus

In a full-term pregnancy (about 40 weeks), the uterus takes in about 600 millilitres of blood each minute. But in a typical delivery, the mother will lose about a quarter of a minute's blood supply – "just" 150 millilitres.

Just think about how amazing this is. The baby comes out, and then the placenta comes out. How come that raw surface, where the placenta used to be, doesn't keep bleeding?

Well, it's partly because of the shape of the arteries inside the uterus. They are spirals. Stretch these spirals out straight, as when the uterus is full size, and blood flows through them easily.

But during the delivery of the little baby, the uterus starts contracting. These contractions squash the uterus – and the spiral arteries which revert to their spiral shape. This naturally closes down the blood flow – resulting in a relatively small loss of blood.

What a lovely piece of Evolution! This clever responsiveness leaves me starstruck.

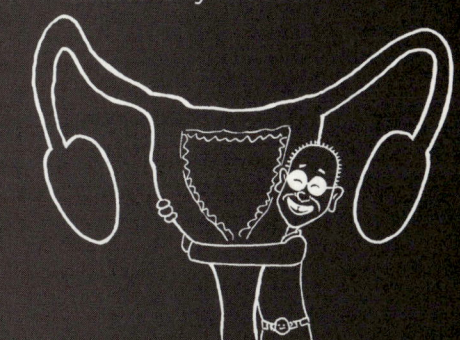

PREMATURITY 101

The Artificial Uterus is not some Frankenstein idea. When you look at the stats for premature births, you understand how useful it could be.

Each year, about 15 million babies are born premature. A baby born before 37 weeks is considered to have arrived early. A full-term pregnancy runs from 37 to 41 weeks.

About 1 million of these premature babies will die before they reach five years of age, due to complications resulting from early birth.

In fact, preterm birth complications are the leading cause of death in children under five years of age. Worldwide, the rate of preterm births varies between 5 to 18 per cent of all births, depending on how wealthy the country is, and how accessible health care is to the population.

Poverty leads to high fatality rates in premature babies. Consider babies born before 28 weeks. If they are born in a wealthy country, the death rate is 10 per cent. Regrettably, in a poor country it's 90 per cent.

But the death rate for premature babies skyrockets if they are born really premature – even in a wealthy country.

About 85 per cent of babies born at 23 weeks or less will die very quickly. Any extra time in the uterus helps survival. So if the delivery can be delayed to 24 or 25 weeks, the death rate drops to 45 per cent, and then 20 per cent.

Premature babies have immature lungs. This is a major cause of their problems – and the major reason for trying to develop an Artificial Uterus.

If you give extremely premature babies oxygen to breathe, you damage their lungs. But if you don't give them oxygen, they will die. This is a problem with no easy solution.

So follow me by looking at a "regular" gestation, and what's inside a "regular" uterus. This will give us an idea of how an Artificial Uterus could help.

> ### Prevent Premature Births
> **The World Health Organization (WHO) has several suggestions to reduce preterm births.**
> One is counselling on healthy diet and optimal nutrition – and avoiding tobacco, alcohol and other substance abuse.
> WHO also recommends a minimum of eight check-ups with a health professional for a pregnant mum. Risk factors for prematurity, such as infections, can be picked up and managed. Checking for multiple pregnancies with ultrasound, and measuring the size of the mother's baby bump, can find other risk factors.

PLACENTA

A lot of stuff has to happen for a fertilised egg to turn into a bouncing baby.

The sperm and egg meet up in the Fallopian Tube, and the fertilised egg migrates down to the uterus. Here it attaches onto the wall of the uterus.

With time, this fertilised egg generates three new biological entities inside the uterus. These are the placenta, the umbilical cord and, of course, the foetus (the cute little bun in the oven). The umbilical cord is the link between the baby and the placenta.

The name "placenta" comes from the Latin for "cake", and from the Greek for "flat, slab-like".

The placenta is a strange organ that is entirely made by the foetus. For both the foetus and the mother, it's simultaneously an essential communication device, and a protective buffer. For the foetus, it's a strange combination of Liver, Lung, Kidney, Gut, and Hormone Generator.

It's a dark-red, round, flat structure. It's about 22 centimetres across, 2 to 2.5 centimetres thick, and weighs about 500 grams. It's a marvel and has many functions. For one, it's a barrier – protecting the mother and the baby from each other. After all, the mother's immune system is always on the lookout for invaders – and half of the baby's DNA is foreign (from the father).

The placenta is also a factory that keeps the foetus happy.

It supplies the growing foetus with everything that it needs to grow into a perfectly formed baby. This includes oxygen, basic nutrition (fats, proteins, carbohydrates, minerals, etc.), hormones and much more. These requirements change with each passing day – and the placenta delivers! It supplies all this essential goodness via the single umbilical vein inside the umbilical cord. Remember that the oxygen diffuses from the mother's blood into the foetus' blood inside the placenta, and then through the umbilical cord directly to where it's needed (the body of the foetus). None of the oxygen enters via the foetus' lungs – they're full of liquid.

The placenta also takes back all the waste products from the baby. Between them, the placenta and the mother "take care" of these waste products for the growing foetus.

The placenta is also called the "after-birth". It's usually naturally expelled about 15 to 30 minutes after the baby has been delivered. Waves of muscular contraction sweep across the uterus and push the placenta out through the birth canal.

You can see how complicated the placenta's job is. Currently, there's no way we can make an Artificial Placenta that can fully mimic a real placenta.

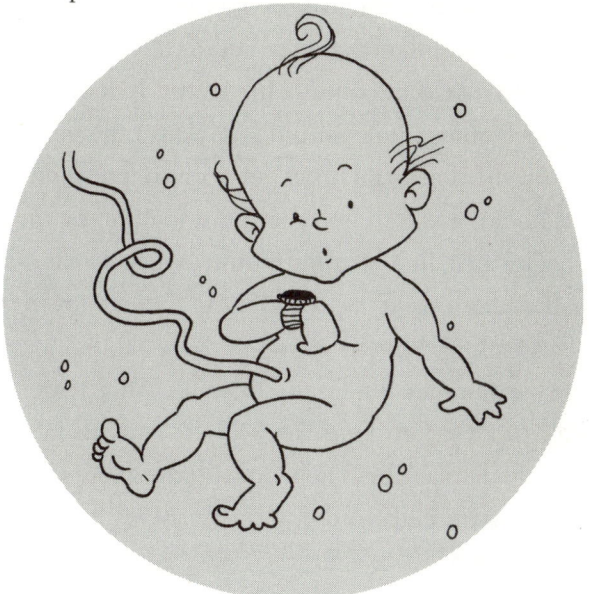

UMBILICAL CORD

The umbilical cord forms around Week 5.

It joins the placenta to the growing foetus. At full term, it carries about 240 millilitres of blood to, and from, the foetus each minute. (This is about one third to one half of the blood flow on the mother's side, inside the uterus). By the end of the pregnancy, it's about 50 centimetres long, and about 2 centimetres in diameter.

It has two arteries (carrying blood from the foetus) and a single vein. Normally, this single vein carries oxygenated and nutrient-rich blood from the placenta, through the future belly button, and directly into the baby's circulation. The two arteries carry the de-oxygenated blood, and waste products, back from the baby to the placenta for further processing.

AMNIOTIC FLUID 101

Now here's something weird.

The foetus' lungs don't deliver oxygen to its blood. They can't, because the lungs are full of a strange liquid called "amniotic fluid". This fluid is both inside and outside the foetus. It does lots of things, besides just cushioning and absorbing shocks for the foetus.

(By the way, the tiny foetal lungs get ready to transition to breathing gas, not liquid, and also become more mature, by hiccuping a lot. They begin their first hiccups around Week 11, and ramp up to about half an hour each day. Read more on hiccups in my 32nd book, *50 Shades of Grey Matter*.)

Early on, the amniotic fluid is mostly water with a few electrolytes (sodium, calcium, etc.). Around Week 13, it's carrying fats, proteins, carbohydrates and urea. These help nourish the baby. There are also various types of stem cells in the amniotic fluid. It is very slightly alkaline, with a pH of 7.0 to 7.5.

One pathway for the amniotic fluid to get into the foetus is through its mouth, and then via the gut wall. But the fluid also goes straight through the skin directly into the blood and tissues of the baby – until around Week 25. At that time, the skin completes its development of an external layer of keratin – and so the skin becomes a lot less permeable. After this, the amniotic fluid is absorbed mostly by the foetal gut.

At first, the amniotic fluid comes entirely from the mother. But around Week 16, the foetal kidneys begin working, so some of the amniotic fluid is urine.

Finally, around Week 40, there's about a litre of amniotic fluid in the uterus.

Swimming in Urine

Hey "ragers", sorry to freak you out. But for about half the time you spent in your mother's uterus, you were swimming in, and drinking, your own urine!

ARTIFICIAL UTERUS

One of the major problems with premature delivery is that the baby's lungs have not developed enough. They are not ready for the transition from "breathing" amniotic fluid to breathing air.

The most recent version of the Artificial Uterus was developed by Dr Alan Flake, a Paediatric and Fetal Surgeon at the Children's Hospital of Philadelphia. He and his team have successfully brought premature foetuses to full term. But we're only talking baby lambs at this stage – not baby humans.

The Artificial Uterus completely bypasses the mother. It does two jobs. It tries to replace the placenta, and it also tries to replace the amniotic fluid. In the Artificial Uterus, the umbilical cord is kept intact. As in the real uterus, it delivers oxygenated, nutrient-rich blood to the baby, and removes the de-oxygenated, nutrient-depleted blood from the baby.

The synthetic amniotic fluid Dr Flake's team used was sterile water, with tiny quantities of sodium, calcium, bicarbonate and more. It had a pH of 7.0 to 7.1, to mimic real amniotic fluid. Unlike real amniotic fluid, which subtly changes during the course of the pregnancy, its constituents were the same for every day of the study. The baby lamb floated in a few litres of fluid, which was constantly infused on one side, and removed on the other side of the Artificial Uterus.

In a real uterus, the umbilical vein and arteries connect to the placenta. In the Artificial Uterus, they connect to a Membrane

Oxygenator, and to a source of nutrition (fats, proteins, carbohydrates, various minerals, plus lots of other chemicals). There was no external pump to move the blood in, and out of, the umbilical cord. Instead, just like in a real uterus, the baby's heart pushed the blood around the circulatory system, and in and out of its body.

In both major components of the Artificial Uterus (synthetic amniotic fluid and oxygenator/nutrition) the artificial version wasn't as good as Nature's version – but it was good enough the keep the lambs alive.

TWO-MINUTE SURGERY

Dr Flake's study involved lamb foetuses at a stage roughly equivalent to 28 weeks for a human baby. However, compared to a human foetus, the body of the lamb foetus was bigger, and the brain was more developed.

It took just two minutes to transfer the baby lamb from the real uterus to the Artificial Uterus. That's 120 seconds from cutting into the umbilical cord to successfully sealing the baby lamb in a polyethylene bag.

Dr Flake did the Caesarean Section by cutting through the pregnant sheep's tummy to expose the uterus, and then cutting again to expose the umbilical cord. He then cut the umbilical cord from the placenta, and instead joined it to a waterproof connector. This connector had pipes leading to the blood-oxygenation device. This oxygenator joined the vein and the two arteries. The oxygenator was connected to a pipe delivering fats, proteins, carbohydrates, various minerals and a few special chemicals. The baby's own heart pumped this oxygenated and nutrient-rich blood into and out of its body.

The baby lamb was then placed into the Biobag – a polyethylene bag with synthetic amniotic fluid. Once sealed, the Biobag with the lamb inside was placed on a heat plate at 39.5°C. Fresh synthetic amniotic fluid was slowly and continuously pumped through the bag.

Over the four weeks of the study, the baby lambs developed well – they grew some wool, had normal sleep/wake cycles, practised normal intermittent breathing and swallowing, and even opened their eyes. Further study showed their bodies, lungs and brains had developed normally. One lamb eventually ended up in a flock, with other sheep.

The altruistic goal of an Artificial Uterus is to reduce the death rate in very premature human babies. It sounds wonderful.

But there are some other critical issues – like how would the parents respond emotionally to seeing their baby floating inside a sealed Biobag for a few weeks? And what about the possibility (a long way into the future) of totally bypassing the natural uterus, and implanting the fertilised egg directly into an Artificial Uterus?

To solve these moral and technical problems, we need not just the brain, but another organ I'm very fond of – the heart.

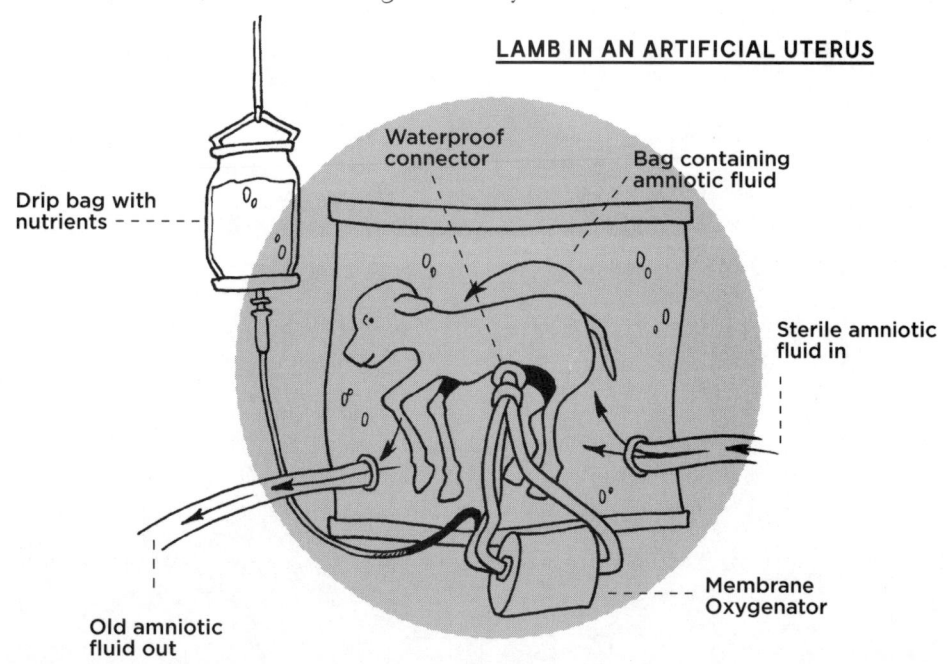

LAMB IN AN ARTIFICIAL UTERUS

05

EARWORM

LET ME REMIND you of Lady Gaga's 2009 hit song "Bad Romance". You probably can't remember all the lyrics but you're probably already humming the very catchy opening. So now I'll apologise for inserting an "earworm" into your brain.

An earworm is a short piece of music that squirms into your brain and won't go away.

EARWORM 101

Earworms are surprisingly common. People can respond to earworms in very similar ways to hearing actual music. The experience can be complex and vivid, and influence your moods and emotions.

A nice description of earworms, back in 1978, was in Desmond Bagley's novel *Flyaway*: "a chant was created in my mind – what the Germans call an 'earworm' – something that goes round and round in your head and you can't get rid of it."

"Earworm" is interchangeable with "Brainworm", "Stuck Song Syndrome" or "Sticky Music". In 2015, the neuroscientist Nicolas Farrugia used the phrase Involuntary Musical Imagery, and defined it as "the non-pathological and everyday experience of having music in one's head, in the absence of an external stimulus".

Nicolas Farrugia's study took 44 people who were prone to earworms. The parts of their brains devoted to sound perception, pitch discrimination, emotions, memory and spontaneous thoughts were found to be larger than average.

MUSIC AND EARWORMS

Some Evolutionary Biologists have claimed that music is an essential step for language development. It turns out that our brains can be kind of pre-wired to embrace music.

Having music in our lives can be useful (as well as beautiful).

Many societies around the world that didn't have written language used music as a way to pass on myths and legends, maps and essential survival tactics. This is the basis of the Indigenous Australian songlines.

Another well-known use of music is the "work song" – to keep the spirits up and to maintain effort. The best example of this is "Heigh-Ho" from *Snow White and the Seven Dwarfs*.

Earworms in Literature

A few science-fiction writers have used earworms as "devices".

In Alfred Bester's book *The Demolished Man*, the hero uses an earworm to stop evil mind readers from being able to read his mind.

Arthur C. Clarke, in the short story "The Ultimate Melody", has his hero try to fashion the earworm to end all earworms. Unfortunately, the hero succeeds, and ends up in a catatonic state from which he never awakes.

WHAT MAKES AN EARWORM

Successful earworms have a few common characteristics.

Firstly, an earworm has to be part of the listener's culture. In general, an earworm won't cross between cultures – so if it works in English-speaking cultures, it probably won't work in Vietnamese or Tibetan cultures. (However, that's not written in stone. I personally am very fond of a few specific Tibetan throat-singing riffs.)

Second, the music is usually a bit faster than average, and often has a certain amount of repetition. The theme from the

Mission: Impossible series of movies is based on a few repeated phrases. Being simple, and having a fairly easy-to-remember melody, also helps the musical piece worm its way into your brain.

Third, it helps if the music is a bit unusual – if it does something unanticipated. This includes unexpected and large jumps in pitch.

TREATMENT OF EARWORMS

Once you've got an earworm, it's kind of hard to get rid of it. An earworm is OK if you like it, but sometimes it gets, well, a little repetitious!

One trick that might work is, rather surprisingly, chewing some gum. It seems that the brain pathways that are used for repeated chewing are also used for repeating music. A 2015 study from the *Quarterly Journal of Experimental Psychology* showed that "chewing gum interferes with voluntary processes such as recollections from verbal memory, the interpretation of ambiguous auditory images, and the scanning of familiar melodies".

Another technique that sometimes works to get rid of an earworm is to play the music over and over until it becomes meaningless. (It doesn't work for me. I have one favourite song that I use for getting me into the mood for writing – and I play it a lot! "Imitation of the Bells" ["*Imitazione delle campane*"] has 8899 plays – and still rising because it has not grown stale. It puts me into an almost hallucinogenic state of mind.)

And sometimes the exact opposite works – go cold turkey and never play the dreaded earworm ever again.

A variation on this is to flood your mind with different songs that aren't as attractive – or just do a task with no background music at all.

If you're flooding your mind with different songs, make sure you avoid the following songs if you don't want to pick up an earworm. One recent study surveyed 3000 people and revealed their favourite earworms. Lady Gaga was Number 1 with "Bad Romance", as well as Numbers 8 and 9 ("Alejandro" and "Poker Face"). In second position was Kylie Minogue with the very appropriately named "Can't Get You Out Of My Head". But, with a name like that, it should have been Number 1 . . .

06

HYDROTHERMAL VENTS AND INVISIBLE MOUNTAIN RANGE

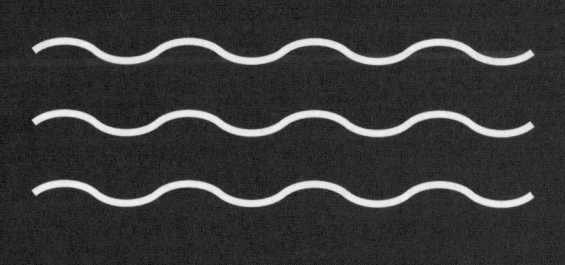

WE'VE FOUND STRANGE forms of Life on Earth. One of the things that makes them odd is that they don't need the Sun to survive. They breed and thrive in what is for us a truly hostile environment. This environment exists in a few hundred spots on the top of the world's longest mountain range – a mountain range that is invisible to practically all of us.

Surprisingly, a new theory claims this hostile environment is the Origin of all Life on Earth.

OCEAN FLOOR LIFE

This Life thrives in tiny isolated communities on the ocean floor – clustered tightly around Hydrothermal Vents. "Hydro" means "water", and "thermal" means "heat" – so a Hydrothermal Vent releases hot water.

We have a few Hydrothermal Vents on land – for example, at Yellowstone National Park in the USA, and on the North Island of New Zealand. They can go under various names such as geysers, hot springs and fumaroles.

But Hydrothermal Vents that are underwater are very different. Overwhelmingly, they sit on the peaks of a huge invisible mountain range. So let's start by looking at where they are.

OCEAN RIDGE

The Ocean Ridge is the biggest mountain range on Earth. But only a few hundred of us will ever see it.

The Ocean Ridge is a virtually continuous underwater mountain range, about 80,000 kilometres long. It winds its way around the Earth, but on the ocean floor. Its shape is similar to the seam on a tennis ball. It can be up to thousands of kilometres wide. This underwater mountain range travels through every ocean basin on Earth.

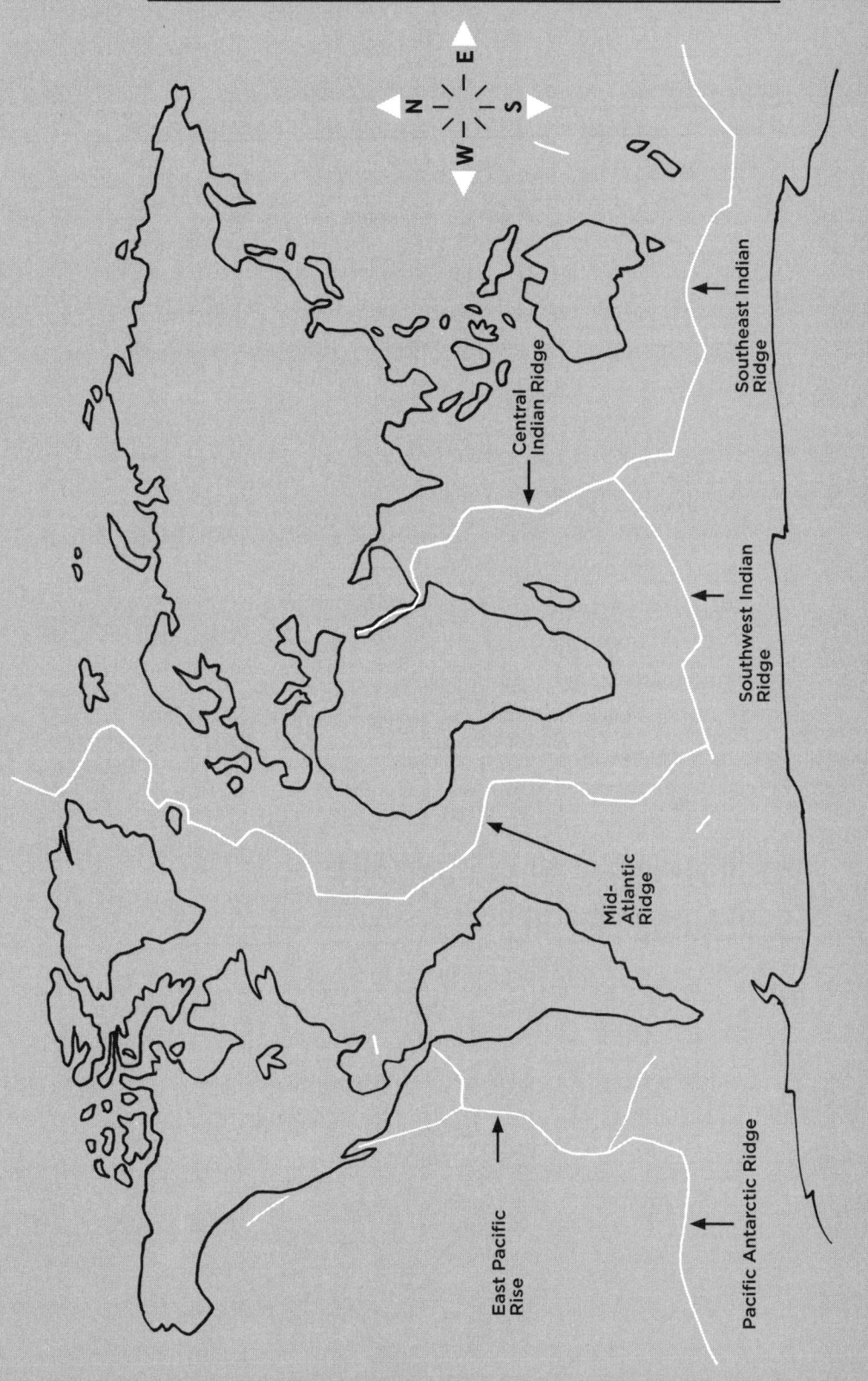

THE EARTH'S OCEAN RIDGE, ABOUT 80,000 KILOMETRES LONG

Back in 1855, Matthew Fontaine Maury of the US Navy proposed the existence of a shallow "middle ground" in his new chart of the Atlantic Ocean. It was originally called the Mid-Atlantic Ridge – because it was close to the middle of the North and South Atlantic Oceans. But later we discovered that some of the ridges (such as the East Pacific Rise) are not in the middle of an ocean. It was only in the 1950s, when the first systematic surveys of the ocean floor began, that the full extent of the Ocean Ridge was proved.

These underwater mountains rise from the depths (about 5 kilometres deep) up to a fairly uniform depth of about 2.6 kilometres below the surface.

The only way you can "see" the Ocean Ridge (as opposed to looking at photos or footage of it) is by going there in a submarine that can dive several kilometres down, and that has special viewing ports. This is a very specialised research vessel, not your average "tourist submarine".

At various locations along the crests of these underwater mountains are regions called "spreading centres". This is where you will find Hydrothermal Vents.

HYDROTHERMAL VENTS ON EARTH – SPREADING CENTRES

Spreading centres run along the top of the Ocean Ridge. They're about 3 to 5 kilometres wide. Spreading centres are where new ocean crust is created. The crust is literally spreading apart, because new semi-molten rock is being pushed up from below and then immediately solidifying. The spreading rate varies between 10 and 160 millimetres per year.

The source of the new crust appears to be vast underground chambers of hot molten magma. This magma rises from these

HYDROTHERMAL VENTS AND INVISIBLE MOUNTAIN RANGE

chambers to the ocean floor. In the East Pacific Rise, these chambers seem to be about 2 kilometres below the ocean floor, about 1 to 4 kilometres across, and about 2 to 6 kilometres thick. Over the whole planet, about 2 to 4 cubic kilometres of brand new ocean crust are formed each year at the spreading centres.

It seems that about 20 per cent of all the heat rising from the core of the Earth comes out through these spreading centres.

The spreading centres have two major regions – a central Neovolcanic Zone, which is surrounded by a Fissure Zone.

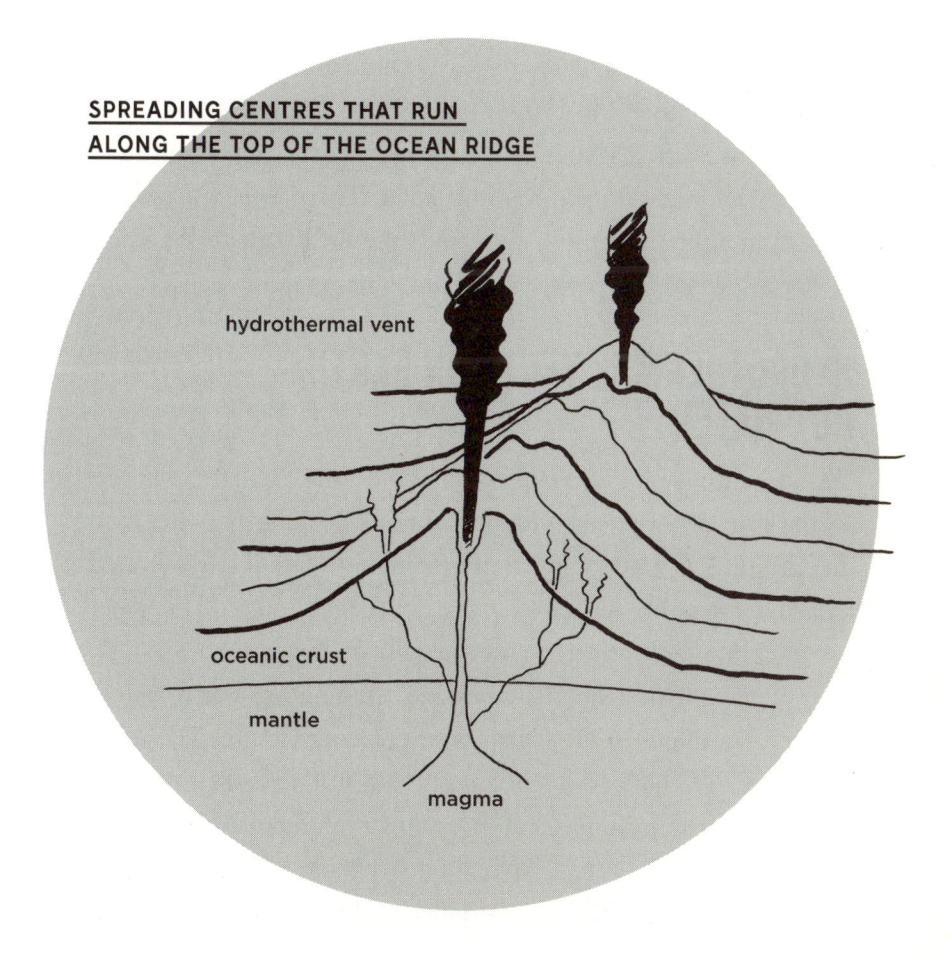

SPREADING CENTRES THAT RUN ALONG THE TOP OF THE OCEAN RIDGE

hydrothermal vent

oceanic crust

mantle

magma

47

The Neovolcanic Zone is about 1 to 2 kilometres wide. This is where the actual spreading happens, caused by the upwelling molten hot magma underneath. This zone can also have active volcanoes and, yes, the fascinating Hydrothermal Vents.

These Hydrothermal Vents have to be immediately above the deeper "bubbles" of hot molten magma. They need the magma to drive the whole cycle.

The Fissure Zone, on each side of the Neovolcanic Zone, is also about 1 to 2 kilometres wide.

The cold ocean water (2 to 4°C) dives into the rock through these fissures. It gets heated, loaded up with minerals, and then squirted out, more alkaline than before, through the Hydrothermal Vents. During this cycle, the elements necessary for Life leach out from the rock, and get dumped into the water. These minerals include potassium, sulfur and phosphorus – all essential to Life As We Know It. There are also iron, copper, zinc and other metals.

HYDROTHERMAL VENTS ON EARTH – HOT WATER

We have discovered hundreds of Hydrothermal Vents – so far.

Back in 1949, scientists found salty water, strangely hotter than the surrounding water, in the Red Sea. In the 1960s, they discovered an active rift in the seafloor, oozing out water at 60°C. This water was uncomfortably warm, but not actually boiling hot.

In 1977, other scientists discovered shimmering warm springs on the ocean floor in the Galapagos Spreading Centre. The water was about 20°C hotter than the surrounding water. To their surprise, they found a strange biological ecosystem around the warm springs.

Prior to 1977, ocean-exploration technology had advanced to include manned submarines and towed underwater sleds with lights and cameras. In 1979 scientists discovered the first really hot Hydrothermal Vents. (I wrote about this in the story "Hot Bugs" in my 3rd book, *Science Bizarre*, back in 1986.)

The water was measured coming out at an astonishing 380°C. (Now we're talking really hot water.) The hot water didn't boil, due to the pressure-cooker effect of a few kilometres of water pressing down – effectively acting as a lid on it. Since then, we have discovered water squirting out of Hydrothermal Vents at temperatures up to 464°C.

As the water squirts out, it can take on many different appearances.

It can be shimmering and fairly clear, or full of bubbles. It can be very dark in colour (the so-called "Black Smokers", usually at higher temperatures) or quite light (the so-called "White Smokers", usually at lower temperatures). The White Smokers expel lighter-coloured minerals, such as calcium, barium and silicon.

HYDROTHERMAL VENTS ON EARTH – LIFE

But most importantly, the expeditions also discovered previously unknown biological communities.

The Ocean Ridge is mostly devoid of Life, apart from around the Hydrothermal Vents.

At the occasional locations where there are Hydrothermal Vents, there are bizarre and very richly dense ecosystems of unexpected life, living entirely off the energy in the hot water. They are the only known life forms on Earth that don't get their energy from the Sun. These isolated ecosystems can measure a few hundred metres in width. The distance between one Hydrothermal Vent ecosystem and the next can be tens, or hundreds, of kilometres. In between the Vents, the ocean floor is in perpetual darkness, with very little life other than bacteria.

Single-celled bacteria and archaea are the basis of this ecosystem that surrounds the life-giving hot-water vents. (Archaea are similar to bacteria, but have different biochemistry.) The bacteria (and archaea) get their energy from strange chemical reactions involving sulfur – not oxygen. (Interestingly, sulfur is immediately above oxygen in the Periodic Table. This implies that they are similar in many ways.)

These bacteria (and archaea) coagulate into thick mats. Then small multicellular creatures such as copepods and amphipods feed directly on the mats.

Bigger creatures such as snails, crabs, clams, limpets, shrimp and fish then eat these smaller creatures. Since the early discoveries, hundreds of completely new species of gastropods and other creatures have been identified at Hydrothermal Vents.

There are also giant tubeworms, 2 metres long and as thick as your arm, and without a mouth or an anus!

How did these bizarre life forms evolve into existence? We don't yet know. In fact, how did Life on Earth evolve into existence? One new hypothesis suggests that Hydrothermal Vents were part of the process.

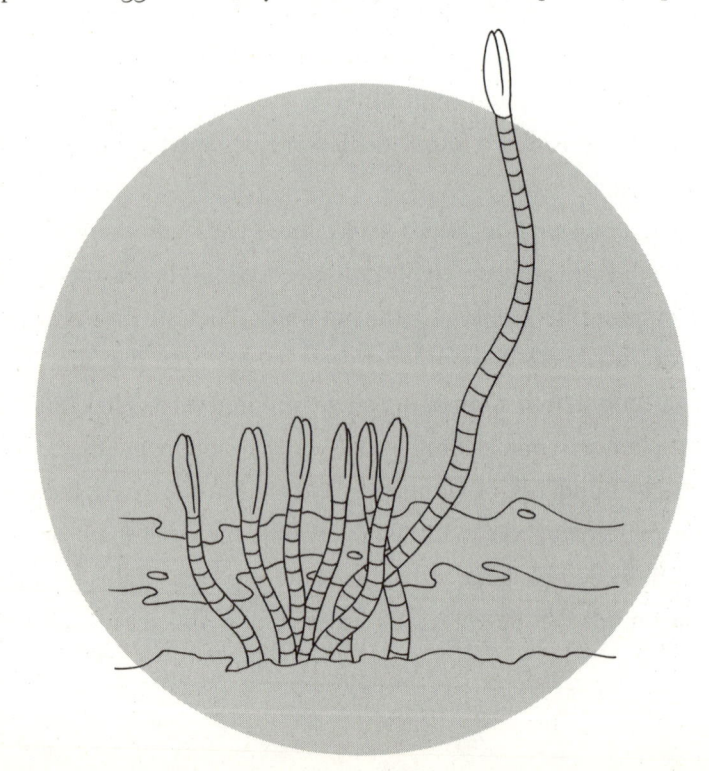

Tubeworms Grow Without Mouth or Anus?!

Tubeworms don't "eat". So how do they thrive and survive?

The tubeworms carry about a billion bacteria in each gram of their tissue. One end of the tubeworm is red-coloured, and full of haemoglobin. The haemoglobin is part of the tubeworm's "energy chain".

Sulfur diffuses out of the hot water into the body of the tubeworm. (That sulfur was originally in the rock, and was dissolved out by the hot water.) The tubeworm's haemoglobin combines with the sulfur, and transfers that energy to the bacteria. The bacteria use that energy to make carbon-based compounds, which they transfer back to the tubeworm. The chemicals needed for growth diffuse into the tubeworm, straight through the skin. And the waste products diffuse out, again through the skin. That's how the tubeworm grows, without having either a mouth or an anus.

It's kind of like if you had a glass panel built into your abdomen. Inside your tummy are lots of bacteria that perform photosynthesis. You go into the sunlight, the bacteria "eat" the sunlight, and pass the energy along to you. You don't need to eat. Same with the tubeworms. But without the sunlight.

Strange Nanorocks

At the Hydrothermal Vents, our scientists have discovered microscopic rock particles, floating in the surrounding ocean. They are generated by the process of hot alkaline water "rubbing" past the rock as it blasts out of the ocean floor.

Amazingly, these microscopic rock nanoparticles are almost identical to the rock nanoparticles discovered in Saturn's rings by the Cassini space probe coming from the proposed Hydrothermal Vents on Enceladus. (See story "Life on Enceladus" on page 54.) There is no other known physical process that can generate nanoparticles of rock in the size range of 4 to 16 nanometres.

Does this mean there are Hydrothermal Vents on Enceladus? Well, if it looks like a duck, and quacks like a duck . . .

ORIGIN OF LIFE?

We still don't know how Life started on Earth.

The much-favoured hypothesis involves the "primordial soup". After the molten Earth cooled down, there were warm ponds of water, scattered across the rocky surface. Somehow an external source of energy (perhaps lightning or ultraviolet light) triggered a series of chemical reactions in this pond. But this energy source would have been intermittent, not continuous.

A new (and still controversial) hypothesis is that Life began around Hydrothermal Vents on the sea floor. Thanks to the molten magma just under the surface, the energy supply would be much more reliable than waiting for a lightning strike, or a break in the clouds so that let the ultraviolet light shine through.

All modern Life uses the energy from food to pump hydrogen ions to a "holding tank" behind a membrane. There's a difference in the levels of hydrogen ions – high on one side of the membrane, low on the other side. When the hydrogen ions flow to the other side of the membrane, they create high-energy chemicals that, in turn, power everything we do.

In the Hydrothermal Vents, there's hot alkaline water flowing upwards and mixing with the more acidic ocean water. This creates a difference in the levels of hydrogen ions remarkably similar to the difference that powers our cells. It might be a coincidence, or it might not . . .

Perhaps the much sought-after Elixir of Life might be as simple as hot water squirting through rock . . .

07

LIFE ON ENCELADUS

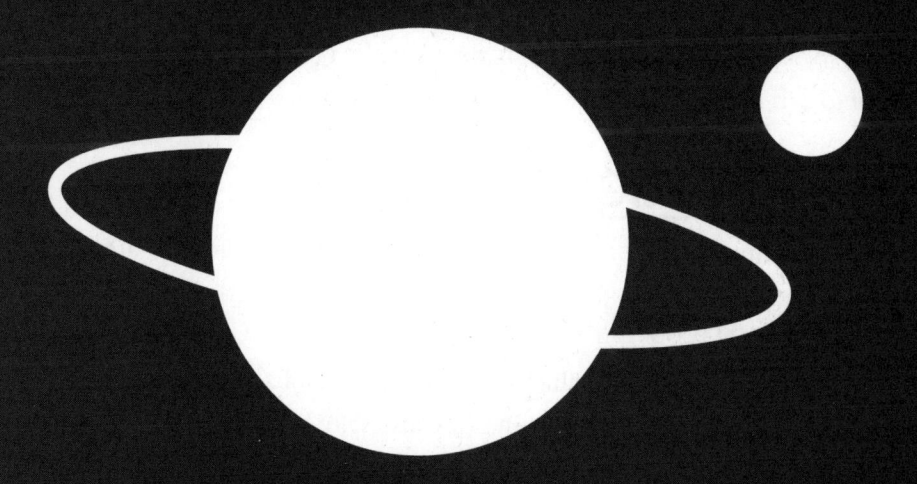

It's not as much fun since Saturn stole my hoop...

THE ONLY LIVING creatures in the Entire Universe are right here, on Planet Earth – as far as we know. But thanks to microscopic rocks, minerals and warm water, we now have a strong suspicion that there might be some Life elsewhere in our Solar System. (At the very least, there are very favourable conditions for Life.)

And where do you reckon that Life might be?

Well, our best bet is a small moon, only about 500 kilometres across. It is called Enceladus, and it orbits the ringed planet, Saturn.

WHY IT MIGHT HAVE LIFE

The argument for Life on Enceladus is that Enceladus shares certain conditions that we have on Earth – and clearly, we have Life on Earth.

First, we're quite sure that Enceladus has a global ocean of salty liquid water. (Earth's many oceans are also salty liquid water.)

Second, we're pretty sure there are Hydrothermal Vents on the rocky sea floor of this ocean on Enceladus. Hydrothermal Vents are fissures in the sea floor with hot water squirting out of them. This water is loaded with the chemicals of Life. (Back on Earth, over the past half-century, we have discovered hundreds of Hydrothermal Vents on the rocky sea floor of our own oceans.)

LIFE ON ENCELADUS

Many of the Hydrothermal Vents on Earth have bizarre biological communities living around them. These truly fantastic life forms found around the deep ocean vents are the only Life on Earth that do not need the Sun for energy. They get all the energy, minerals and nutrients they need from the hot water squirting out of the Hydrothermal Vents. (To read more about this, check out "Hydrothermal Vents and Invisible Mountain Range" on page 42.)

Life flourishes around Earth's Hydrothermal Vents – and it doesn't need the Sun to survive.

So, likewise, there might be Life clustered around the Hydrothermal Vents on Enceladus.

Let's go into a little more detail – but don't worry, unlike some Big Life Discussions, this one's not too heavy.

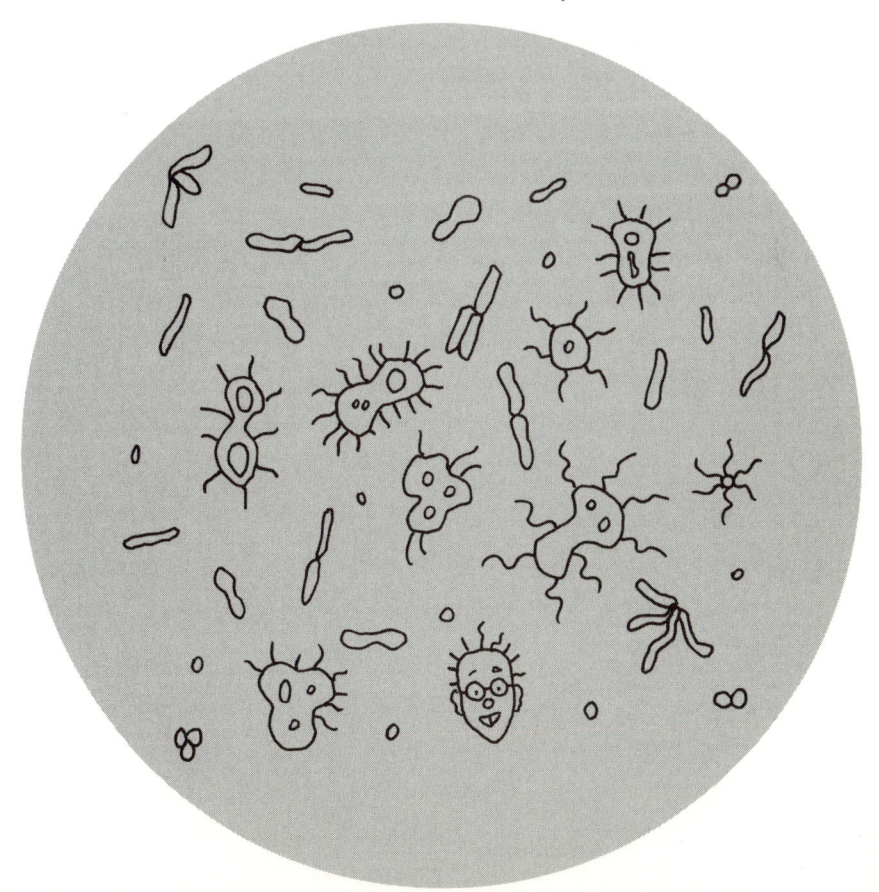

57

ENCELADUS 101

Enceladus is cold, which you would expect, because it's a long way from the Sun. The surface temperature of Enceladus is down around −200°C. It's the sixth biggest of the 62-plus moons of Saturn. It orbits Saturn once every 33 hours.

Enceladus has three distinct layers.

At the centre is a small rocky core. The rocky core is about 400 kilometres across. The rock is important for the possibility of Life – because it's naturally loaded with the elements of Life. These elements (sodium, nitrogen, phosphorus, etc.) can dissolve from the rock into the seawater to give a lovely chemical soup. These elements can combine in many different chemical reactions to make the building blocks of Life.

(It's lucky that the centre of Enceladus is not just water. If you just have pure H_2O and no other elements besides hydrogen and oxygen, then Life is very unlikely.)

The layer surrounding the rocky core is a global ocean of liquid water – that's right, friendly water, not some other "hostile" liquid such as ammonia or methane. This ocean has an average depth of about 26 to 31 kilometres – deeper at the poles, shallower at the equator. (The oceans on Earth have an average depth of just 3.7 kilometres.)

Floating freely on top of the ocean, totally disconnected from the rocky core, is a shell of ice. Again, fortunately for possible Life, the ice is made from water, not something inhospitable such as carbon dioxide or ammonia. This ice varies from about 5 kilometres thick at the poles to about 35 kilometres thick at the equator. (Yes, the ice thickness runs opposite to the water thickness.)

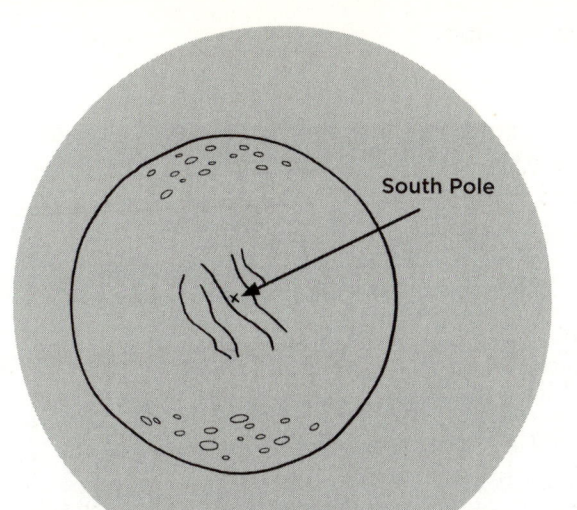

TIGER STRIPES ON ENCELADUS'S SOUTH POLE

THE ICE PLUMES OF ENCELADUS

The South Pole of Enceladus has four "cracks" or fissures in the ice called "Tiger Stripes" – because that's what they look like. They're huge – ranging from 100 to 300 kilometres long. These fissures open up when Enceladus is further from Saturn, and close down when it's nearer.

Amazingly, back in 2005 we discovered plumes of frozen water squirting out of these Tiger Stripes. The water in the ocean is a liquid. But somewhere on its 5 kilometre journey outwards from the ocean, through the cracks in the ice called the Tiger Stripes, and finally into the vacuum of space, it gets frozen from liquid water to solid ice. The plumes are about four times brighter when the Tiger Stripes open up. Sometimes there can be 100 plumes erupting at the same time.

Scientists eventually found all kinds of chemicals in the plumes – sodium, potassium, carbon dioxide, propane, acetylene, formaldehyde, methane, molecular nitrogen, and more. All these chemicals can be associated with Life.

On average, about 200 kilograms of ice jets out every second – and at speeds of around 2000 kilometres per hour.

> ### Will Plumes Run Enceladus Dry?
>
> There are many plumes of ice squirting out of Enceladus. Will they use up all the water?
>
> Not for a long time. At the current rate, by the time our Sun becomes a Red Giant star in 5 to 6 billion years, only 30 per cent of the total mass of Enceladus will have been used up.

CASSINI AND OCEAN

Back in 2005, we didn't know where this frozen water was coming from. A tiny reservoir of water, or a huge ocean?

The answer turned out to be a global ocean.

We got this answer from hard-won information gathered from a few dozen flybys by the Cassini space probe, a joint endeavour of NASA, the European Space Agency and the Italian Space Agency. These flybys took place over more than a decade. (Yes, exploring space is a long game.)

Cassini is the second-largest unmanned space probe ever launched – 5.7 tonnes at launch, 6.8 metres long, and with 14 kilometres of internal wiring. It was launched in 1997, and arrived at Saturn in 2004. (That's right, Saturn is so far away that Cassini took 7 years to get there.) Cassini then inserted itself into orbit about Saturn, to zip repeatedly through the mini solar system that is Saturn and all its 62-plus moons.

It's not easy to prove the existence of a hidden ocean under the ice. For one thing, the measurements taken have to be so exquisitely sensitive that they can pick up changes as small as one part in a billion.

While flying closer than 100 kilometres from the surface of Enceladus, Cassini broadcast radio signals to NASA's Deep Space Network.

LIFE ON ENCELADUS

The power of the radio was tiny – only 81 Watts! That's less power than is needed to run a car's headlights. The scientists analysed the frequency of these radio signals, using the same scientific principle that Police Radar uses to measure the speed of cars – the Doppler Principle. They used it to work out Cassini's speed.

> I just love the Deep Space Network webpage: http://eyes.nasa.gov/dsn/dsn.html. You can see data coming in and control signals going out. This site tells us what they are doing with the US$18 billion of assets that NASA is using to explore our microscopic region of the Milky Way.

GLOBAL OCEAN

The scientists found that Cassini, depending on whether it was approaching or departing Enceladus, would speed up (or slow down) by a few millimetres per second. The Cassini scientists could measure changes in velocity as small as 0.090 millimetres per second – so a few millimetres per second was almost too easy. (Depending on what scientists were trying to measure, Cassini would fly past Enceladus at speeds between 31,000 and 65,000 kilometres per hour.)

Of course, the scientists had to compensate for any other influences that could possibly speed up or slow down Cassini. These included the tiny pressure of sunlight from the very distant Sun, the drag (or friction) from the ice in the plumes as it collided with the space probe, the heat radiating from its nuclear-powered electrical generator, and so on.

Using the changes in speed of Cassini as it flew past Enceladus, they worked out the moon's exact gravitational field – and how it varied with distance away from its core. These variations were due to different densities of different types of matter, at different depths under the icy surface.

After three flybys by Cassini, the scientists had enough data to get the answer as to where the frozen water in the plumes was coming from.

Enceladus has an ocean, made of water. The new question was – how big is this ocean?

The Cassini scientists also saw that Enceladus has a strange wobble.

Taking the wobble into account helped them work out that Enceladus has a very big ocean – in fact, a global ocean. This ocean completely surrounds the rocky core. And in turn, this ocean is also completely surrounded by a floating cover of ice.

The ocean is the middle of a strange "sandwich" – it's the filling between a rocky core and a frozen outer shell made of ice.

WHY IS ENCELADUS'S OCEAN WARM?

But there was another mystery.

Saturn is so far from the Sun that very little heat reaches the planet. The Sun is only 1 per cent as bright as it is on Earth. How come the water is warm enough to be liquid, i.e. above 0°C?

Even today, we're not 100 per cent sure why it's above freezing.

But we think a large part of the answer is the process called "tidal heating". Basically, Enceladus is constantly being squeezed this way and that – thanks to Saturn's massive gravity. Friction generates heat.

Enceladus is subject to much greater distorting forces than our moon. Saturn is about 95 times heavier than Earth – so it has lots more gravity. To make the distorting forces even stronger, Enceladus is closer to its massive parent, Saturn, than our moon is to the Earth – only 238,000 kilometres away, versus 384,000 kilometres for our moon.

Thanks to Saturn's huge gravity, and its closeness, the sphere of Enceladus is always being massaged and kneaded. These movements create friction, which continually heats the ocean water under its icy covering.

Of course, there are other potential factors that can also generate heat. These include the decay of radioactive elements in the core of Enceladus, chemical reactions between the rock and the liquid water, and so on. So we haven't fully solved the problem of where all the heat is coming from.

However, we are quite sure there's a global ocean of liquid water on Enceladus. And on this ocean floor, we think there are Hydrothermal Vents.

HYDROTHERMAL VENTS: ROCK

What's the evidence for the Hydrothermal Vents on the rocky sea floor of Enceladus?

On Earth, we've discovered nanoparticles of rock coming out of Hydrothermal Vents. These nanoparticles can be formed only under very specific conditions.

In 2004, Cassini had discovered almost identical microscopic rock particles orbiting in the E Ring around Saturn.

To be specific, the scientists noticed that as Cassini inserted itself into orbit around Saturn, it ran into a rain of microscopic buckshot.

But back then, Cassini had bigger targets to examine (such as the fabulous rings of Saturn) and the job of dropping a probe onto the gas-covered surface of Titan, the biggest moon in the Solar System. The nanoparticles information was filed away for later study.

When the scientists did analyse the nanoparticles, they found that they were rock (virtually pure silicon dioxide). But, very strangely, these tiny rocks covered only a very narrow range of sizes – about 4 to 16 nanometres. (A nanometre is a billionth of a metre.)

There's only one way we know that Mother Nature can generate rock particles in this narrow range of sizes. It's to have very hot, alkaline water gushing out through fissures in rock.

So these tiny rocks are the reason to believe that Hydrothermal Vents exist on Enceladus.

Sample Return?

So far, we humans have had only a handful of Sample Return missions from outside our planet.

Besides the 381 kilograms of Moon rocks from the Apollo missions, we have had a few other Sample Return missions – Stardust, Genesis and Hayabusa. Stardust returned samples of dust from Comet Wild 2, Genesis gave us some samples of the Solar Wind, while Hayabusa returned samples of Asteroid 25143 Itokawa. The Russian Luna programme also returned 326 grams of Moon rocks.

It would be very easy to get samples of what is squirting out of the 90-odd plumes of ice erupting from the South Polar region of Enceladus. No landing or drilling is needed.

All we have to do is fly through a plume of icy water.

HYDROTHERMAL VENTS: FROZEN WATER

There was a huge flurry of excitement after the Cassini Orbiter had successfully inserted itself into orbit around Saturn, then started examining the rings and even soft-landed a probe onto the land-and-liquid surface of Titan. But it made more great discoveries over the next decade.

In 2005, it photographed the plumes spurting out through crevasses in the thin ice at the South Pole of Enceladus. On one flyby it even zipped right through an erupting plume of icy water. (The ice particles were about one micron across. A micron is a millionth of a metre. For comparison, red blood cells are about 7 microns across, while human hair is about 50 to 70 microns thick.)

The instruments on Cassini "tasted" this ice. The water ice carried salt (identical to the salt, or sodium chloride, in our oceans on Earth), but was slightly more alkaline than our Earthly oceans. (I wrote about "Ocean Acidification" in my 34th book, *Game of Knowns*.) The erupting ice also carried organic molecules, such as methane and carbon dioxide, formaldehyde, propane and acetylene, as well as biologically available nitrogen in ammonia.

HYDROTHERMAL VENTS: HYDROGEN

In 2017, NASA announced that Cassini had discovered hydrogen in the erupting plumes of frozen water.

Cassini had used its instruments in a more sensitive mode, by flying past at only 31,000 kilometres per hour, and at an altitude of less than 50 kilometres above the surface. It found that the Enceladus plumes are about 98 per cent H_2O, about 1 per cent hydrogen and about 1 per cent ammonia, with trace amounts of carbon dioxide, methane and other chemicals making up the rest.

If the chemistry on Enceladus is similar to what happens on

Earth, the hydrogen was almost certainly produced by hydrothermal reactions between the hot rocks and the water in the ocean of Enceladus.

Hydrothermal Vents on Earth spew forth hydrogen. This hydrogen directly supports rich microbial life. It's their energy supply. Our Earth microbes need only 10 parts per million of hydrogen to serve as an energy source. Microbes use the hydrogen to reduce carbon dioxide, to make methane (methanogenesis).

Cassini found relatively high levels of hydrogen gas in the plumes. So, that's plenty of hydrogen as an energy supply for bacteria on Enceladus to potentially get started.

FINALLY, LIFE ON ENCELADUS?

It seems that every single one of the ingredients needed for ocean Life on Earth are also present on Enceladus!

The stage is set for the possibility of Life – but we still have no hard data. So our space missions are moving from general exploration to more focused astrobiology missions. The plumes on Enceladus are the best astrobiology targets in the entire Solar System.

Perhaps the Hydrothermal Vents on the tiny moon of Enceladus could play host to the hottest event in our history – the discovery of Extraterrestrial Life!

(So here it comes again . . .)

Perhaps the much sought-after Elixir of Life might be as simple as hot water squirting through rock . . .

Shiny Enceladus Made E Ring

Saturn, the ringed planet, has many rings. Overall, the rings are only about 10 metres thick, and made of lumps that range in size from a speck of talcum powder to a small apartment.

The E Ring is the widest and second outermost ring – but also the most diffuse. Enceladus orbits right in the densest part of it. The E Ring is unstable – its lifespan is between 10,000 and 1,000,000 years. Left alone, it would quickly fade away and entirely vanish. So if it's still here then it has to be constantly replenished. We now know that the E Ring is "restocked" by the plumes squirting out through cracks and crevasses at the South Pole of Enceladus.

However, some of the ice in the plumes doesn't make it into the E Ring. Instead, it falls back down to the surface of Enceladus as a kind of "snow" – up to 100 metres thick. This makes part of the surface of Enceladus very "new", and also, very reflective.

In fact, Enceladus is the most reflective body in the Solar System. Because it's so reflective, it's colder than the other moons of Saturn.

What Kind of Life on Enceladus?

Just to spell it out, this next section is 100 per cent Pure Speculation. I reckon (with zero proof) that we will find Life on Enceladus

at the bacterial level – single-celled creatures.
I further reckon (again, with zero proof) that
we will find multicellular Life – fish and the like.

Would that Life share the same origin as our Life on Earth?

An easy way to check would be to look at the "chirality", or handedness, of the chemicals that make up this Life. On Earth, practically all Life has so-called "right-handed" sugars, and "left-handed" proteins. (I discussed this in the story "Left versus Right" in my 12th book, *Sensational Moments in Science*.)

If we find the same handedness, then Life on both Earth and Enceladus could share a common origin. But if we find the opposite on Enceladus, that very strongly implies that we have found a second, and totally independent, Genesis of Life in our Solar System. That would be a soft argument for Life in other solar systems.

Life Elsewhere in Solar System

The prime candidates for Extraterrestrial Life in our Solar System are Enceladus, Europa, Mars and Titan. (I love the topic of Extraterrestrial Life in Space. I wrote about Life on Europa, and Life on Titan, in my 18th book, *Munching Maggots, Noah's Flood and TV Heart Attacks,* **and Life in Deep Space in my 4th book,** *Science Bizarre II***.)**

Enceladus has the four essential ingredients for

Life As We Know It on Earth. It has liquid water, energy, carbon and nitrogen. Nitrogen is essential to make amino acids and proteins. But we don't know how long the ice has been liquid for. Even 100 million years is a short time for Life to evolve.

Then there's Europa – a moon of Jupiter. It's similar to Enceladus, with an extensive global ocean of water under an icy covering. It has also a few plumes shooting out frozen water. So far, we've found no hint of the minerals or chemicals of Life in its scant icy plumes. (This might be related to not having had a long-term orbiter in the Jupiter system. But the NASA space probe Juno arrived in mid-2016. Who knows what it will find?)

Mars seems to have had both an atmosphere and liquid water in the past – but hardly any is left now. It is also short in nitrogen. But at the very least, I reckon that we'll find Life as bacteria, underground at the poles, where the ice is frozen water. At some depth, the temperature will rise enough for the ice to turn into liquid water.

Titan is intriguing. It has rain, rivers and oceans of a liquid – but the liquid is methane, not water. The atmosphere is only 1.5 times more dense than Earth's atmosphere. Titan's atmosphere is (like Earth's) about 80 per cent nitrogen – but while our remaining 20 per cent is mostly oxygen, Titan has 20 per cent methane. Finally, Titan is a lot colder – around −180°C.

If there is Life on Titan, it might be quite different from what we know as Life on Earth.

08
REFRIED BEANS

IT SEEMS ONLY fair to expect that Refried Beans are beans that have been fried more than once. But no. The Northern Mexican dish has been fried only once.

What!?!

The completely misleading name of this dish comes from a mistranslation from the Spanish.

The dish we call *"frijoles refritos"* is based on beans. It can be used in other meals (such as tostadas) or rolled into a tortilla to make a Bean Burrito, or eaten with corn tortilla chips.

THREE STAGES

The beans go through three stages of preparation.

First, they are infused with water. This is done by very slowly soaking them in water overnight, or more quickly by simmering or stewing them on the stove, or cooking them really quickly in a pressure cooker. Second, they are mashed into a paste. Finally, they are fried until they have lost lots of moisture – often in a meat-based fat such as lard or bacon fat (for extra flavour). Confusingly, they are sometimes neither fried nor baked, and simply served after being mashed.

LANGUAGE ISN'T AS SIMPLE AS ABC

The Spanish word for "beans" is "*frijoles*", and the "*frito*" in "*refritos*" means "fried".

So "refritos" should mean "re-fried", shouldn't it?

After all, in English, placing the prefix "re-" in front of a verb means that we carry out that action more than once. You can "reheat" your leftovers, or "refocus" your attention, or even "retrain" someone to give them a new field of knowledge.

But in the Spanish language, the prefix "re-" has a different meaning. It means "very".

So if something is very good, it is "*retebien*", while to burn cooking oil is "*requemar*". So "*refrito*" means well fried (until they are more dry), not fried twice. But if you ask for a refill of refritos, you'll get a second serve.

09

MISOPHONIA - HATING SOUND

THE HUMAN BODY is so incredibly complex. We're still learning about it. In fact, we are still discovering new diseases.

So let me introduce you to a disease so contemporary that you won't always find it in your spell checker, or your dictionary (if you've still got one). Indeed, this disease is so modern it might not even turn out to be a real disease (which will make it incredibly on-trend in this alt-truth world).

Let's dive in to "misophonia" . . .

MISOPHONIA 101

The word "misophonia" literally means "hatred of sound".

Misophonia is a disorder of how the brain processes sounds.

The unfortunate sufferer gets extremely powerful negative reactions in response to perfectly ordinary, everyday sounds.

These "Trigger Sounds" are overwhelmingly generated by fellow humans. About 80 per cent of them are related to the mouth and nose, while about 60 per cent have a strong repetitive element. Trigger Sounds include hand noises (such as typing or clicking a pen), eating, chewing, footsteps, breathing, yawning, and even the plosive sound of the letter "P", as in "Peter Piper picked a peck of pickled peppers."

Other non-human Trigger Sounds include the humming of a refrigerator, the knocking of water pipes, the ticking of a clock or a cat licking its paws.

Doctor with Misophonia

A doctor who both has misophonia and treats it, wrote about his strange predicament in *The New York Times*.

Having misophonia meant that noises made by his patients set off very negative reactions in his brain – even if they merely yawned. But on the other hand, his professional role could override his emotional response. He was able to recognise that his primary duty was the care of his patient.

What a strange battle must be going on in his brain with regards to his attitude to his patients. Perhaps a combination of acute annoyance (because he has misophonia) and intense empathy (because he has misophonia).

RAGE RESPONSE

In people with misophonia, the response to the Trigger Sound is extremely emotional. It's a bizarre and extreme combination of rage, terror, fear, panic and anger. It seems to be related to the Flight or Fright (or Freeze) Sympathetic Nervous System response (which evolved to get you ready for dealing with emergencies).

In about 5 per cent of cases, hearing the Trigger Sound leads to the sufferer committing actual physical violence. In about 25 per cent of cases it results in verbal violence.

In the remaining three quarters of cases the anger gets bottled up. Mostly, the symptoms are mild to moderate – contrary to many media reports. But in some cases, misophonia can interfere with day-to-day life. Misophonia can stop sufferers doing their regular work, making new friends or interacting with the friends they already have.

Here are some of the stories from psychiatrist George Bruxner's research and *The New York Times*.

A mother couldn't stand the noise of her daughter eating carrots. Her reaction was very physical. She tried to be calm – but failed. She rapidly progressed from deep breathing and a racing heart to agitated pacing, and even grabbing at her own hair. Unfortunately, she ended up shouting and swearing at her own daughter, totally out of control.

Another typical trigger was related to random noises in public places. Sufferers had great difficulty with travelling on public transport or going to the movies. They couldn't stand the sounds of people sniffling, clipping their nails, or even random mouth noises. They had to move away.

One sufferer had a very visceral reaction to sounds such as slurping soup or snapping lips together. She felt as if she had been physically hit. If she used all her willpower, she could tolerate the noise for just a few moments. Then she had to run away.

This shows how the reactions caused by misophonia can strongly interfere with the sufferer's daily activities.

Most people with misophonia have insight into their condition. They intellectually realise that their aggressive reaction to a perfectly innocent Trigger Sound is both excessive and unreasonable, and that the loss of self-control is socially unacceptable. So they often try to avoid their Trigger Sounds by all kinds of workarounds, including wearing noise-reduction headsets. Unfortunately, these headsets cut out a whole bunch of other necessary, and useful, sounds as well!

History of Misophonia

Misophonia was first identified in 1997 by the audiologist Marsha Johnson. She called it "4S" or "Selective Sound Sensitivity Syndrome". It was the researchers Margaret and Pawel Jastreboff from Emory University who first coined the actual term "misophonia", around the year 2000.

Even as recently as 2013, only a few case studies had been published.

This condition has not yet been accepted into the Psychiatric Bible known as the DSM-5 (*The Diagnostic and Statistical Manual of Mental Disorders, Fifth Edition*).

Misophonia Is Not . . .

Misophonia is quite different from hyperacusis, where an ordinary sound appears to be excessively loud and painful.

It's also very different from tinnitus, which involves a ringing sound in the ears. (As an aside, tinnitus actually "happens" in the brain rather than in the ears.)

Misophonia is not an Obsessive-Compulsive, nor an Impulse Control, Disorder. Furthermore, it's not an eating disorder – but in some cases, it might turn out to be related (for example, where a person can't bear the sound of their own teeth chewing on food).

WHAT'S HAPPENING?

On average, misophonia appears around age 12, but it can turn up over a fairly wide range of ages. Unfortunately, it worsens with time – in other words, you get sensitised to a wider range of Trigger Sounds.

A relatively small study in early 2017 looked at 20 people with misophonia. It showed that the Trigger Sounds set off changes in heart rate and a change in electrical skin resistance, due to sudden "micro-sweating".

This study also looked at brain activity with misophonia.

First, it showed that people with misophonia had abnormal activity in the Anterior Insular Cortex when exposed to their Trigger Sounds. The Anterior Insular Cortex tells our "consciousness" what it should pay attention to. It integrates internal inputs from the body with external sensory inputs. It also processes emotions.

Second, upon hearing the Trigger Sounds, the brain centres that process sound fired up much more strongly than in people without misophonia. This sounds reasonable for a disorder of the auditory system.

Finally, the brain scans also showed a remarkable degree of hyperconnectivity between the auditory systems and the systems that process emotion.

It's early days so far (especially for a condition that is not yet even officially recognised), but there are some treatments that seem to work.

Both Cognitive Behaviour Therapy and some brain-training techniques appear to have a degree of success – sometimes as high as 83 per cent. Alcohol seems to make misophonia better, while caffeine can make it worse. (Mmm. See my story on how Alcohol can make you more deaf called "Alcohol and Hearing" in my

40th book, *The Doctor*. Maybe there's a connection?) Check out the Interwebs for more specific information.

But any treatment is better than nothing, when an innocent yawn can trigger a misophoniac's bite . . .

10

SHOUTING AT HARD DRIVES

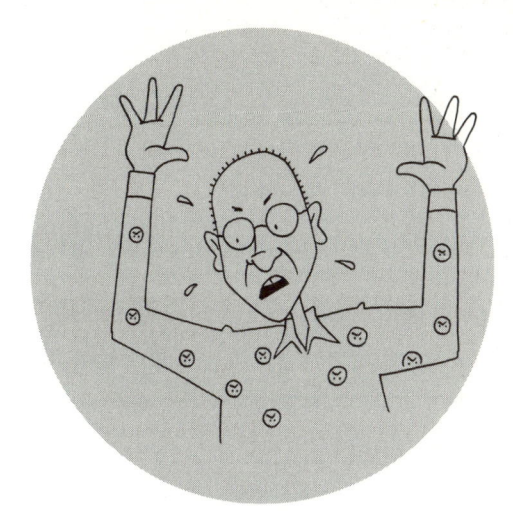

EVER SHOUTED AT your computer, in sheer frustration, because it was running so slowly? Unfortunately, it might make your computer run more slowly!

(And no, this wasn't because you hurt its feelings. Although you did shake it up!)

RUST VERSUS ROCK

Ironically, it's the hard drive of the computer that can be affected by shouting. They're not as tough as they sound.

Hard drives store information, even when the power is switched off. There are two types of hard drives today. They are based on quite different materials – Rust or Rock. (OK, I am being a bit metaphorical here, but there is a lot of literal truth as well.)

"Rust" is found inside spinning hard drives, while "Rock" is in the faster (and more expensive) Solid State Drives (SSDs).

By Rust, I mean iron. (OK, actually "iron oxide", but I'm taking liberties here in the name of alliteration.) This type of hard drive has a spinning disc (called a platter), covered with a layer of material (basically iron) that can be repeatedly magnetised or demagnetised. A moving mechanical arm sweeps a tiny magnetic head across the platter, between its centre and perimeter. As the magnetic head moves, it can write or read information. This information appears as tiny magnetic fields in the iron oxide.

Spinning hard drives are cheap, but they read and write information fairly slowly. Because they're cheap, lots of computers (and most data centres) still have spinning hard drives. Unfortunately, these have lots of moving parts. This means spinning hard drives don't like being dropped, shaken, stirred – or even shouted at.

PARTS OF A SPINNING HARD DRIVE

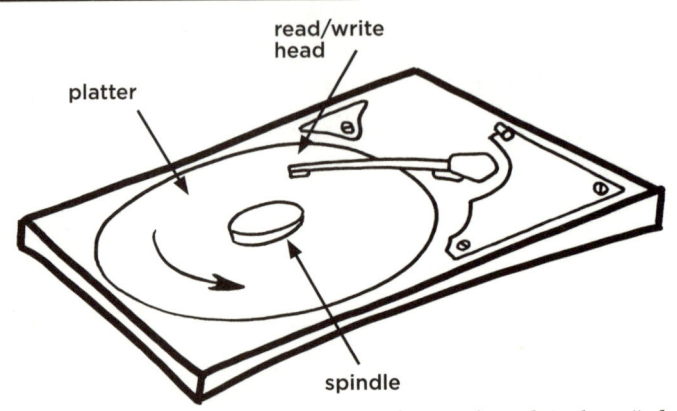

[Diagram showing a spinning hard drive with labels: platter, read/write head, spindle]

By Rock, I mean silicon. (OK, "silicon dioxide", but "silicon" is pretty close.) This type of hard drive stores information in ultra-purified silicon. This is the Solid State Drive (or SSD).

SSDs are more robust, because they have no moving parts. They also use less power, and are quieter, faster and less likely to fail (immune to shouting) – but they are more expensive. However, the price of SSDs is dropping rapidly.

HARD DRIVE 101

The very first spinning hard drives were introduced by IBM in 1956. Since then, the improvements have been enormous.

Their volume has shrunk by a factor of 56,000 – from bigger than a double fridge to the size of a pack of playing cards. The cost to store data has plummeted by a factor of 300 million. The amount of data these hard drives can store has jumped from 3.75 megabytes to 10 terabytes – almost 3 million times more.

The modern spinning hard drive carries probably the finest micro-engineering you can buy cheaply. It's built at the scale of billionths of a metre, or nanometres.

The Rust covering the spinning platter is very thin (just 15 nanometres thick). The tiny magnetic head that sweeps across the spinning disc can accelerate at up to the equivalent of 500 Gravities. A pilot will usually become unconscious after experiencing 5 Gravities for 5 seconds.

The magnetic head that does the reading and writing is about 10 nanometres thick, and about 100 nanometres wide (about one ten-thousandth of a millimetre across). It hovers above a disc that is spinning at up to 15,000 revolutions per minute, and at a height of some 4 nanometres. It's held up by a tiny bubble of air. If the magnetic head were to push through that bubble of air and actually touch the spinning disc, it would spew up zillions of tiny particles – and cause immediate catastrophic failure of the hard drive.

There are about 10,000 parallel magnetic tracks in each millimetre of that spinning disc. If the head wanders away from the centre of a magnetic track by more than 25 nanometres, it cannot read or write data.

READ/WRITE HEAD

Data bits that have just been written

> ### Multiply by 1 Million
>
> **The sizes inside a hard drive are so small they are hard to appreciate. So let's multiply everything by a factor of a million.**
>
> The magnetic head that does the reading and writing is now expanded to about 10 millimetres thick, and about 100 millimetres wide. (It's about the size of a skinny wallet.) It hovers above a disc that is spinning at up to 226 million kilometres per hour (a fifth of the speed of light!), and at a height of some 4 millimetres.
>
> The read/write head is held up by that 4-millimetre bubble of air. There's not much margin for error.

FIRE SUPPRESSION

The "Shouting at Hard Drives" story begins with the first known fire in a computer data centre. It happened on 2 July 1959 in a computer room operated by the United States Air Force inside the Pentagon. It burned for over five hours, and caused about US$7 million in damage. It was started by a light bulb.

Very quickly, the National Fire Protection Association in the USA brought in NFPA Standard 75 to deal with computer-room fires.

Clearly, it's not good to pour water onto electronics.

So what do you use to put out a fire in a data centre full of spinning hard drives, if not water?

It makes sense to deprive the fire of oxygen by flooding the room with a non-flammable gas, from pressurised cylinders.

Today, we use Inert Gas Fire Suppression Systems. (The inert gas is usually a mixture of nitrogen, argon and carbon dioxide.) The gas is stored inside cylinders at pressures up to 300 times greater than atmospheric pressure. It floods into the fire zone and dilutes the oxygen level from around 20 per cent to around 13 per cent. Combustion needs at least 16 per cent oxygen – and so the fire goes out. Luckily, people can still function at 16 per cent oxygen.

But there were teething problems. The noise of the inert gas escaping through tiny nozzles can be very loud. It can be loud enough to kill the spinning hard drives – causing more damage than the fire!

How loud? Well, in some cases, as loud as a passenger jet taking off right next to you.

SOUND KILLS HARD DRIVES

Back around 2007, we began to have reports of fire-extinguishing systems causing hard-drive failures.

There have been documented failures in Australia, France and Romania, when the noise of the fire-suppression system killed hard drives in the data centres of big banks. Data centres use spinning hard drives, because they are cheaper.

To put some numbers on it, 110 decibels, about as loud as a petrol-driven chainsaw 1 metre away from you, will degrade the performance of most hard drives. Sounds that are louder than 130 decibels, similar to a trumpet blast at half a metre, will totally stop most spinning hard drives from delivering data.

Legal standards for fire alarms specify sound levels typically between 90 and 120 decibels. So sometimes just the fire alarms themselves can kill the hard drives.

In general, if the noise level is harmful to humans, it's harmful to computers.

COSTLY CURE?

The cure is easy, but a little expensive.

The cylinders of inert gas are now being fitted with more nozzles, so there's less gas moving through each spout. The nozzles themselves are being shifted further away from the hard drives, and are no longer aimed directly at them. Some nozzles have been redesigned to greatly reduce the noise (by 20 decibels), without affecting the flow rate of the inert gas. The hard-drive cabinets are being equipped with sound-insulating doors. And of course, the spinning hard drives can be shut down. There's also a shift to the more expensive SSDs, which have no moving parts.

SHOUTING AT HARD DRIVES

How did we discover that just shouting at a spinning hard drive can very significantly degrade its performance? Back in 2008, Brendan Gregg, an Australian computer engineer, did an experiment while he was working for the computer company Sun. Watch the video at https://www.youtube.com/watch?v=tDacjrSCeq4 – it shows him in a noisy room full of spinning hard drives. He then places his mouth very close to one bank of hard drives, and shouts/roars very loudly. Suddenly, the meters show that the data-transfer rate from these hard drives drops precipitously!

Most probably, the loud sound of his shouting sent vibrations through the delicate innards of the hard drives. (In a similar way, you've probably felt your chest vibrate at a loud concert.) Think back to the tiny head sitting a tiny distance above the speeding platter. In the design, everything has been optimised for maximum efficiency.

Any change whatsoever makes the data transfer slower. This change can be the head sitting a bit closer or further away from the platter, or not placed directly above the track.

In other words, any kind of misalignment of the read/write heads to the data tracks and platter will slow down the data-transfer rate. Yelling can be enough to throw them out of line.

Maybe there's a lesson in all of this. When you are frustrated, either by Life in General or a stubborn computer, don't shout – just whistle a cheerful song. Quietly.

Personal Anecdote

I was flying to Perth from Sydney. The weather was bad, so the pilot had climbed a little higher than normal. This meant the air pressure inside the cabin was quite low.

The cabin crew came around with dinner. Being ever so keen to please, I quickly lifted my laptop so they could serve my dinner. This relatively minor, but abrupt, motion was unfortunately enough to make my hard drive's read/write head push through the tiny bubble of air holding it above the spinning platter.

To my horror, multiple lines appeared on the laptop screen as the hard drive died. Boo hoo. I had to have wine with that meal!

11

PONYTAIL SWING

WHEN YOU RUN or jog, you don't just move forward. You actually move in three directions. You run forward, your head bobs up and down a little, and there's a tiny amount of side-to-side motion.

But here's the weird bit. If you have a ponytail, when you run it mostly moves side to side! Why? It turns out that in this special case, any sideways motion, no matter how minuscule, is amplified exponentially. The "special case" is that in a runner, there's a mathematical coincidence between the natural frequency of a foot hitting the ground, and the natural frequency of a ponytail swing.

The Mystery of the Swinging Ponytail was tricky to solve. The answer involved a lot of Maths, a bit of Astronomy, the 2012 Ig Nobel Prize, and one of the great mathematicians of the last century, Joseph B. Keller.

HAIR 101

Even the great Leonardo da Vinci himself studied hair (a tradition proudly upheld by the Shampoo Laboratories of the advertising world, where incredibly Beautiful People do their Science in ultra-white labcoats). Leonardo wondered why falling hair took on the same shape as water plunging downwards in a waterfall. He wrote in his notebooks how "hair shafts" looked like the "streamlines" in water.

It took until 1946 before scientist C.M. van Wyk was able to formulate the first known Equation of State for hair, by studying wool. (Yes, Physicists have great curiosity about everything in the world around them – and Maths is the tool they use to solve problems.)

On one hand, a ponytail looks like a simple and well-behaved object. But in reality, it has many separate components – lots of hair shafts. Each hair exerts forces on its neighbours. (These forces are not trivial for hair care. Think of how much emphasis your favourite shampoo puts on "volumising" your hair.)

You have around 50,000 to 100,000 individual hair shafts growing out of your scalp. These individual shafts have their own shapes and stiffness, and they collide with other hairs when they get pulled into a ponytail. So an Equation of State for hair would involve factors such as the weight of each hair shaft (which is about 65 micrograms per centimetre, "or in more amusing units, 6.5 grams per kilometer"), elasticity, and the Radial Swelling Force. The Radial Swelling Force? Well, in a ponytail, the individual hairs are confined, or held in place, by neighbouring shafts of hair – and they react to this force by trying to swell radially outwards.

Van Wyk's Equation of State was the first step but it needed work. Work takes time. This work took over half a century.

Finally in 2012, the paper "Shape of a Ponytail and the Statistical Physics of Hair Fiber Bundles" was written by Raymond E. Goldstein and his team. The authors devised the whimsically named "Rapunzel Number". This Number was essential to understanding how "after washing and rinsing, the hairs in a drying ponytail pass through a glass transition with decreasing humidity". A glass transition is the reversible transition from "hard and brittle" to "viscous and rubbery".

The authors were awarded the 2012 Ig Nobel Prize in Physics for this work.

The Ig Nobel Prize is given to "honour achievements that first make people laugh, and then make them think". (I was awarded an Ig Nobel Prize for my groundbreaking research into Belly Button Fluff, and why it is almost always blue. I have a story on the 2016 Ig Nobel Prize on page 254.)

Equation of State?

"Equations of State" are everywhere. They are used to design wings of jet planes, work out how to melt chocolate and have it retain its texture when it cools, and understand how geckos stick to ceilings.

It is an equation (or bunch of equations) that describes how something physical behaves.

That "something" could be the plasma inside a star, or the Thermohaline Ocean Current (which travels through the oceans of the world, and moves huge chunks of water – the size of continents – every second), or the compressed air inside your bicycle tyre.

So with the air in your bike tyre, an Equation of State might take into account the air's temperature, pressure, volume and internal energy. In this case, you might use this Equation of State, known as the Ideal Gas Law:

$$pV = nRT$$

where p = pressure
  V = Volume
  n = number of moles of air
  R = Ideal Gas Constant
  T = Temperature

MATHEMATICIAN FOR ALL SEASONS, AND REASONS

But also sharing the hair-themed 2012 Ig Nobel Prize in Physics was the great mathematician Joseph B. Keller. His contribution was solving the mysterious sideways motion of a jogger's ponytail.

Keller saw mathematics in everything. He was awarded many honours, including the prestigious Wolf Prize in Mathematics, and the National Medal of Science.

He had worked out how to reduce the dripping of tea from a teapot — simply pour the tea out more rapidly. This is easier to do if you only partially fill the teapot. (He had previously shared the 1999 Ig Nobel Prize in Physics for this essential research.)

Keller had used the Theory of Queues to explain why adding just a few extra airline flights per hour can cause massive delays. He also showed us how to avoid delays.

He calculated why an earthworm can easily move across your favourite fluffy bedspread, while a snake cannot. The snake has a backbone and needs friction to move forward — which it can't find on a fluffy bedspread. But an earthworm has no backbone. It can move forward just by sending waves of expansion and contraction along its body.

Keller also did lots of work for the military.

During World War II, he developed his Geometrical Theory of Diffraction. It explored how to use underwater sonar to find both submarines and explosive mines. More importantly, his Theory showed how waves (whether acoustic or electromagnetic) could bounce off objects — and even bend around corners.

He then used this knowledge to design antennas for sonar and radar. He even showed how to shape a vehicle to avoid detection, and become "invisible" — which today we call Stealth Technology.

But he also did maths just for fun – such as finding out why ponytails swing from side to side.

To solve the Ponytail Problem, he had to go back to the work of the American astronomer George William Hill.

Hill's work didn't start off looking at hair. Instead, he was trying to work out if the orbit of the Moon around the Earth was stable. And so in 1886, he derived what is now known as Hill's Equation. But totally accidentally, Hill's Equation was essential to understanding the sideways movement of a ponytail.

PONYTAIL SWINGS – THE OLD ONE-TWO

So let's think about the ponytail as being a swinging pendulum. Of course, it has its own natural frequency. An average ponytail is around 15 centimetres long, and so it naturally swings at around 1.4 cycles per second. (The maths says so – look up "Pendulum" on Wikipedia.)

But the ponytail is supported by the head, which is joined to the runner's body. The body's frequency depends on how speedy the runner is. An elite runner will have a natural cadence of around 170 to 190 steps each minute. But the average runner will have a slower natural rate of around 140 to 160 steps each minute. So for a runner, hitting the ground 168 times per minute, the head is moving up and down at about 2.8 cycles per second.

By a lovely coincidence, the natural frequency of the ponytail's movement is about half the natural frequency of the ponytail's support – the runner's head. The sideways motion happens only because of this coincidence.

And now for the Big Finish.

Hill's Equation tells us that if there is this 1:2 relationship in frequency between a pendulum and its support, then any sideways motion at all (no matter how small) will be amplified exponentially. (I can't explain this in words. You have to use the language we call Mathematics.)

For a runner, there is indeed this 1:2 relationship between the natural frequencies of the runner's ponytail and their head. And yes, there is a tiny sideways motion – because the runner hits the ground with the left foot, and then the right foot, and so on.

And so, thanks to Hill's Equation (and Keller's Hairy Application of it), we now know why the tiny initial sideways motion of the ponytail is amplified until it reaches its own maximum rhythm – as seen on runner's ponytails, all around the World.

I love ponytails. Knowing why they swing from side to side just makes them mathematically more beautiful.

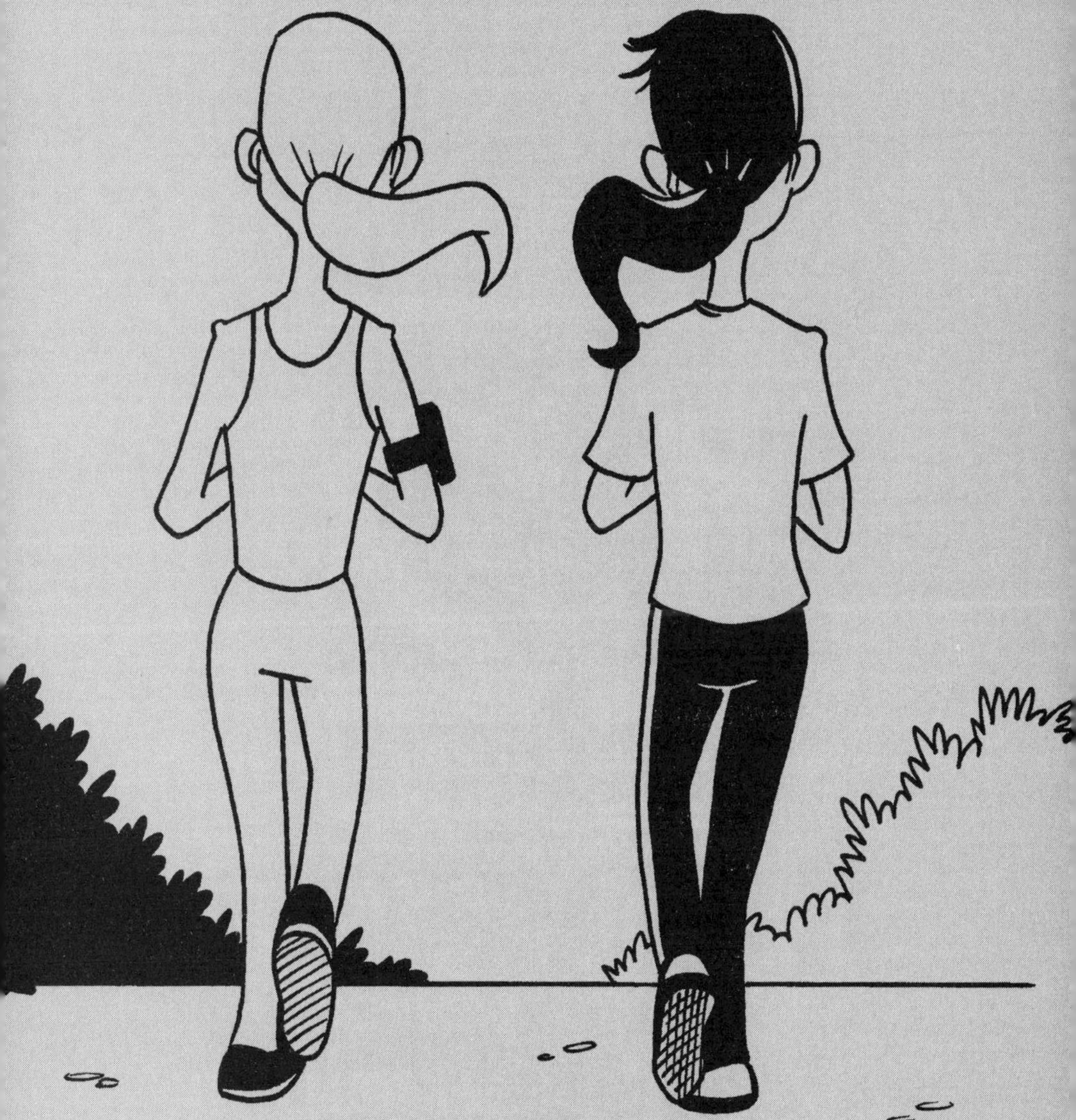

12

LIGHTNING POWER

LIGHTNING IS SO impressive that when it happens, it simply stops you in your tracks. It looks incandescent and powerful. And so, loads of people have wondered, "Why don't we capture the electrifying energy of lightning bolts, and use it to power our industries, and boil our kettles?"

The surprising answer is that there's not enough energy for us to even bother catching it – compared to how much energy we use just to run our society.

LIGHTNING 101

Across the whole planet, lightning flashes into existence about 1.4 billion times each year. Most of this lightning happens in the Tropics, drops off in frequency further from the Equator, and doesn't happen at all at the North and South Poles.

One and a half billion lightning bolts sounds like a lot, but only about a quarter of it gets to the ground. The rest of the lightning happens either within a single cloud (Intra-Cloud lightning) or from one cloud to another (Cloud-to-Cloud lightning).

Even today, there are many mysteries about lightning. For one thing, we don't fully understand how the bolt of lightning gets started. We have measured the electric fields inside thunderstorms over several decades. This hasn't helped us answer the question of how lightning starts. Unfortunately, these electric fields are simply nowhere near big enough to set off that first spark. Maybe Cosmic Rays from outer space create a cascade of electrons, which then increase the local electrical field. Maybe some ice crystals have positive charges on one side, and negative on the other, which again can increase the local electrical field. But at this stage, they are all still hypotheses – we simply do not know.

(I have written about lightning in several of my books. There are stories on lightning, and ball lightning, in my 8th book, *Latest Great*

Moments in Science. I wrote about Red Sprites and Blue Jets in my 13th book, *Pigeon Poo, the Universe and Car Paint*.)

> ## Lightning Deaths
>
> In 2016, lightning killed about 38 people in the USA. The US lifetime odds of being hit are about 1 in 3000. The number should unfortunately climb in coming years, with further turbulent weather and storms due to more heat energy in the atmosphere from Global Warming.
>
> In India, about 1000 people are killed by lightning each year. Many more survive, but are left with severe neurological injuries.

CAUSE OF LIGHTNING

We have worked out bits of the lightning puzzle. Even though we don't know how lightning gets started, we do know the two essential elements for lightning to happen. These two elements are located inside a thunderstorm.

The first is air moving rapidly upwards. The second is freezing cold (temperatures ranging between 15°C and 25°C below zero).

These rather perverse conditions produce three types of water. These three types of water are necessary for lightning to happen.

First, and rather surprisingly, there's some actual liquid water. It's in the form of small water droplets that are supercooled, and so remain liquid (even though the temperature is way below zero).

Second, as you would expect, there are small crystals of hail – solid frozen ice.

Third, and again a bit unexpectedly, there are larger (and heavier) blobs of soft hail, with the unusual name of "graupel". Graupel is about 2 to 5 millimetres across, and happens when supercooled water freezes on the outside of (say) a snowflake.

So now we have all three types of water, let's start the machine running.

Thanks to the strong updrafts of wind, the graupel runs into the other two types of water (the small supercooled droplets and the small ice hail). Via a complicated process, the liquid and the small hail become positively charged, while the graupel becomes negatively charged.

The supercooled water droplets and hail are much smaller and lighter, so the strong up-currents of wind inside the thundercloud carry them skywards. This makes the top of the thundercloud become positively charged. The positive charges repel each other, and so they spread out away from each other. This is part of the reason that you get the so-called "anvil" – the large flattened top of a thundercloud.

But the third type of water, the graupel, is larger and heavier, and either stays where it is in the strong updraft, or falls to the bottom of the cloud. Either way, the bottom of the cloud becomes negatively charged. In turn, this negative charge on the bottom of the cloud induces a positive charge on the ground, a kilometre or more below.

THUNDERCLOUD PREPARING FOR LIGHTNING

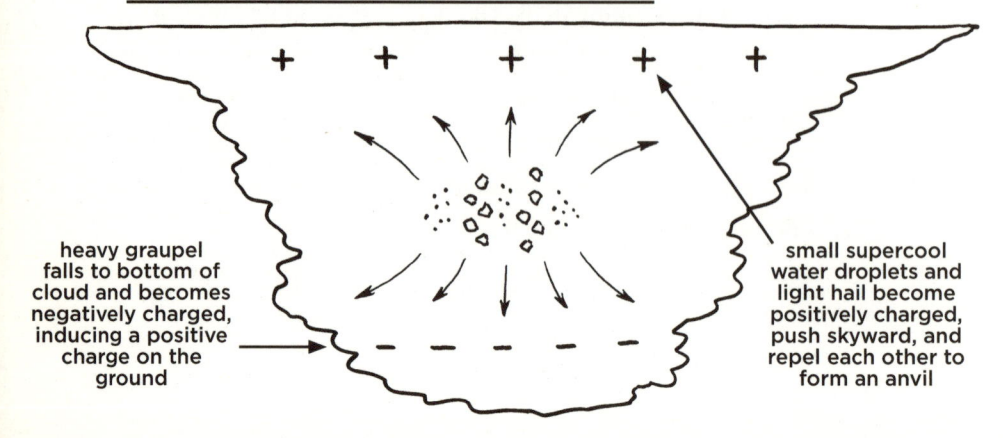

heavy graupel falls to bottom of cloud and becomes negatively charged, inducing a positive charge on the ground

small supercool water droplets and light hail become positively charged, push skyward, and repel each other to form an anvil

Scary Power of Lightning

One stormy afternoon, my family and I were going for a walk along the cliffs before the rain hit. Suddenly, my daughter said, "Hey look, Mum's hair is lifting up." Sure enough, on both my wife and my daughter, tendrils of hair were lifting up from the scalp. It wasn't a full complete fan of hair, with every single hair shaft standing up – like you would get from a Van de Graaf generator. But lots of hair shafts were standing up. Sadly, with very little hair on my scalp, I had to content myself with looking at the back of my forearm, where the hairs were standing up as well.

Obviously, we were inside a powerful electrical field – and we were all charged. I wasn't too worried about being hit by lightning (we didn't have the full fan of erect hair). But as a drama queen, I was a tiny bit worried.

Suddenly, there was a flash of light from a cloud a bit away from us. At that exact instant, everybody's hair fell down. The electrical field had been discharged to zero. There was no lightning bolt to the ground. It happened entirely within the cloud that lit up.

Nine seconds later, I heard the rumble of thunder. So that meant that the lightning bolt was 3 kilometres away. That was amazing! Even from 3 kilometres away, the charge building up was enough to lift hairs away from the scalp.

A few seconds later, tendrils of head hair were again lifting up on my wife and daughter.

Intrepidly, we kept on walking – despite the storm. After all, the lightning was kilometres away.

THE SPARK HAPPENS

This electrical tension between the bottom of the cloud and the ground can be huge – in extreme cases, it can make all your hair stand on end, like a fan.

Eventually, taking about a fiftieth of a second, a lightning strike stutters all the way down from the bottom of the cloud towards the ground in a series of steps. Each step is about 45 metres long – a tad smaller than an Olympic swimming pool. This downward-going lightning bolt is clearly visible – even though it's travelling about 200 kilometres per second.

The downstroke carries about 10 to 100 amps of current. (For comparison, 10 amps running a domestic electrical heater will make the power cable a bit warm, while 100 amps is a bit more than your entire house can draw from the electrical grid.)

Now this downstroke – impressive as it is – is just the warm-up for the Main Event. On its way down, the downstroke has built up a conductive pathway between the bottom of the cloud and the ground. What happens next is the Return Stroke – which is much more powerful, yet kind of invisible.

POWERFUL INVISIBLE RETURN STROKE

More powerful? Sure.

Instead of carrying 10 to 100 amps of current, it carries 30,000 amps – and sometimes as much as 120,000. This stupendous amount of current is enough to create temperatures 10 times hotter than the surface of the Sun – 50,000°C. This incredible temperature creates the characteristic blue-white colour of a lightning bolt. The air heats up almost instantaneously, and expands explosively, producing the shockwave that we hear as thunder.

So is it more powerful, but "kind of invisible"? Yes.

It's kind of invisible because it's moving so fast. While the downstroke of the lightning strike moves at about 200 kilometres per second, the upward Return Stroke zips along at 100,000 kilometres per second – about one third of the speed of light. All you see with the naked eye is the slower downward strike and a really bright flash of light for the upward Return Stroke.

The only way you can "see" the upward movement of the Return Stroke is if you use a Military-Industrial-Scientific–grade Ultra-High-Speed camera. With this, you can capture a few frames of a very thick and very bright upward-going stroke. But otherwise, all you see is a generalised flash. It blends into the other flash – the downstroke. The naked eye certainly can't see it going upwards.

LIGHTNING CAPITAL OF WORLD

Where should we go to build our first pilot plant to try capturing this enormous amount of potential energy?

The obvious location is what NASA (after 16 years of satellite monitoring) declared to be the Lightning Capital of the World. It's on Lake Maracaibo in the state of Zulia in Venezuela. While Venezuela is in the continent of South America, it's so far north that it's actually in the Northern Hemisphere – running from the Equator to about 10 degrees North.

In Lake Maracaibo, lightning storms happen about 297 days of each year – that's over 80 per cent of the time. In fact, the locals have a lightning bolt on the official state flag – to better fit in with the scenery.

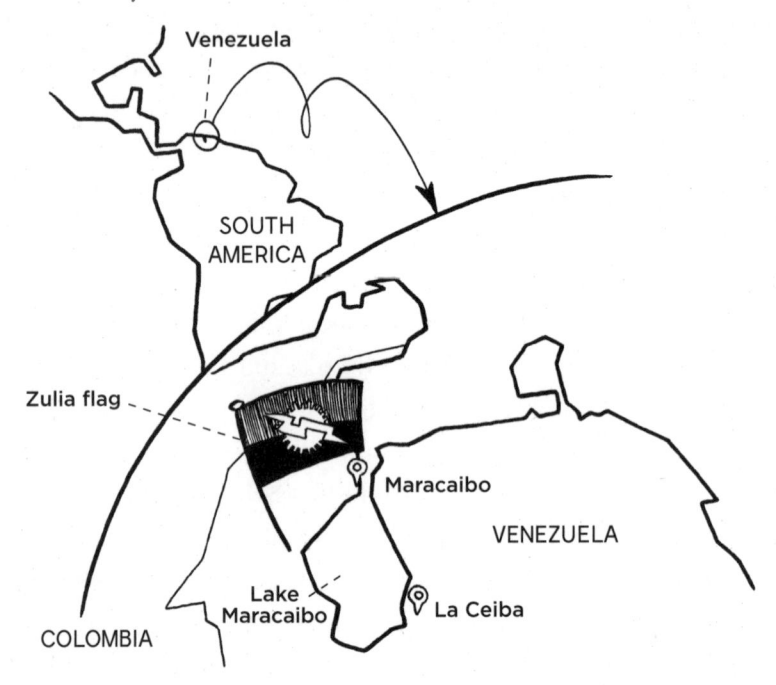

Now "Lake" Maracaibo is not actually a lake – because it's connected to the sea by a strait. (But if the locals call it a lake, I'm happy to call it a lake. After all, it is the largest "lake" in South America.) The lake is about 200 kilometres across, and is a major shipping route for Venezuela's crude oil.

So keeping it straight, we've got the lake (which is not really a lake), in a part of South America (that happens to be in the Northern Hemisphere) – and which happens to be the Lightning Capital of the World.

WHAT MAKES A LIGHTNING CAPITAL?

Lightning storms are so constant on one part of this "lake" that they've earned the name "The Never-Ending Storm of Catatumbo". The name comes from the Catatumbo River that empties into Lake Maracaibo. The water is warm, and the atmosphere is very humid – after all, it's only 10 degrees from the Equator. The mouth of the river is surrounded on three sides, like a horseshoe, by three mountain ranges.

When the cold, dry air from the mountains above meets the hot and humid air down below, you've got the best possible conditions for lightning. The storm clouds build up to an altitude of over a kilometre. Within an hour of the storm clouds forming, the lightning starts flashing. The flash rate can reach 200 flashes per minute. A few people in the region get hit by lightning every year.

A typical lightning storm lasts for 10 hours – and this happens nearly 300 nights each year. The clouds are like an enormous diffuse light bulb, flashing in the sky. It's bright enough to read a newspaper – in the middle of the night. The storms reach their peak in September but according to the locals, the prettiest storms happen in November each year.

These storms are so powerful, and so regular, that they have been used by European navigators for the last four centuries. In fact, they're nicknamed "Maracaibo's Lighthouse".

COLD HARD NUMBERS

So we've come to the right place to develop the technology to capture and use lightning. But before we start thinking about building lightning rods and enormous banks of giant ultracapacitors, let's take a look at what the scientists would call "The Numbers". To be specific, let's look at how much energy we could capture, to see if it's even worth doing.

Typically, each lightning bolt carries about 500 megajoules (MJ) of energy. What does that mean in plain English?

First, 500 MJ is the amount of energy needed to run an average Western house for about a week. Second, 500 MJ is roughly the amount of energy in 14 litres of petrol or gasoline. And third, 500 MJ is enough energy to bring about 1500 kettles of water to the boil. (That's assuming that your kettle holds one litre of water.)

Suppose that we could capture all the energy from all the 1.4 billion lightning bolts that happen each year. In that case, we would have enough energy to heat up 280 kettles each year for each person on the planet. If each litre makes four cups of tea, that works out to each person getting three cuppas each day.

Now that's quite surprising. Before I did The Numbers, my (incorrect) Gut Feeling was that the energy from lightning could easily provide enough energy to run the whole world.

Instead, all that the world's lightning could do is give everybody a few cups of tea each day.

Even though lightning is very impressive, it's no equal to the energy-hungry society that we humans have developed over the last few centuries. With no trouble at all, we can burn more than 14 litres of petrol in commuting to work for a week – and that's the amount of energy in just one lightning bolt.

So harnessing lightning can't compete with fossil fuels or renewables – but it's still enough to boil up the billy, and light up the sky.

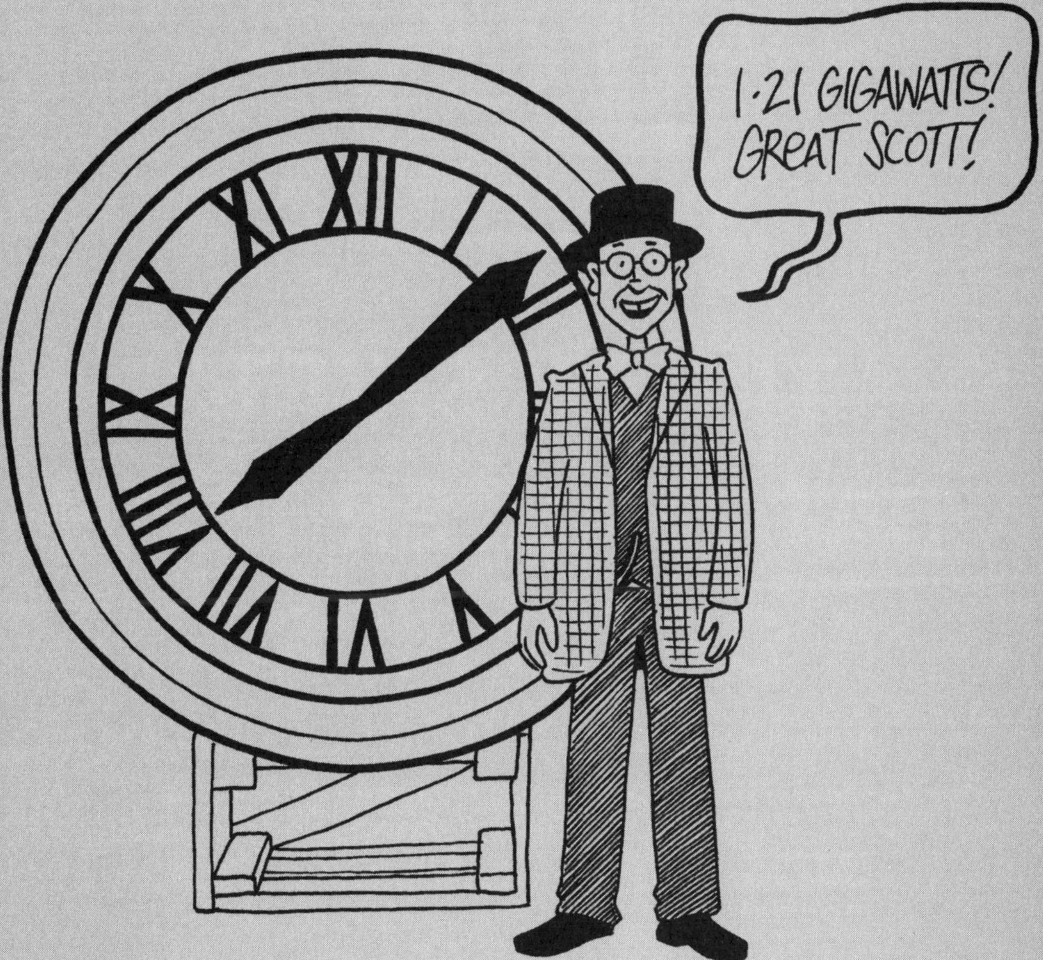
1·21 GIGAWATTS! GREAT SCOTT!

13

TRUCK SHIFTS DATA FASTER THAN OPTIC FIBRE

EVEN THOUGH OPTIC fibre is blindingly fast, if you want to move huge amounts of data quickly, what you need is a big truck. Yes, a truck.

A truck can shift data faster than optic fibre – if you're specifically trying to load 100,000 terabytes of data onto the cloud.

So if you're an image company like DigitalGlobe (DGI), you've specifically got about 100,000 terabytes of satellite images that you've accumulated since 1992. Now, DGI doesn't know which particular image the next customer will want. It might be a picture of a village on a river (for a government campaign about flood relief), or maybe tree cover changing with time. DGI has to store all the images. Even if the customer only wants a 1 megabyte picture, DGI still has to have 100,000 terabytes of images in the cloud, ready just in case.

Now if you've got optic fibre and a 1 gigabit per second connection (about 100 times faster than the average Australian internet connection speed) it will take about 30 years to upload 100,000 terabytes. For comparison, your personal computer probably has between 0.5 terabytes and 6 terabytes of storage.

AMAZON SHIFTS BITS, AS WELL AS BOOKS

So what do you do? Thirty years is too long to wait.

Well, Amazon first got into the Data Shifting business in 2015 with "Snowball". It's a box a bit bigger than the cabin luggage you can take on a domestic flight. It is packed full of hard drives, and can store up to 80 terabytes of information. The data is encrypted for security as it goes into Snowball. The customer has the only key. (I wrote about Public and Private Key Encryption in the story "Bitcoin: Legend of a Ledger" in my 40th book, *The Doctor*.)

Snowball turned out to be so popular that by 2016, it had snowballed into the Snowmobile.

Snowmobile is an 18-wheeler semitrailer truck with a 15-metre-long shipping container in the back. This container carries over 1250 Snowballs, giving a total capacity of around 100,000 terabytes. The container has a fire-suppression system, water inlets, and a 350-kilowatt power supply providing cooling and power to all the individual hard drives. There are multiple 40 gigabit per second connections configured to handle 1 terabit per second. At this speed, the 100,000 terabyte storage could be filled in just a few weeks.

The truck can travel right across the USA in under four days – giving an effective data-transfer speed of around 2500 gigabits per second from coast to coast. That's about 2500 times faster than optic fibre. Even if you include the time needed to write as well as read the 100,000 terabytes, plus transport time driving to a cloud data centre, the data-transfer speed of the Snowmobile is still about 500 times faster than optic fibre – just several weeks versus 30 years.

Of course, data is valuable, so in addition to the encryption, Amazon has gone Old School with their cybersecurity, and is happy to rent you a dedicated security escort vehicle.

ONLY CHANGE IS PERMANENT

We are temporarily in a period where physical transport of huge amounts of data is faster than internet transport. It's reminiscent of the old "sneaker-net". This was when it was quicker to load the information onto a memory stick, and walk in your sneakers to a computer on the other side of the office, than to transfer it online.

But electronic transport is going to have to get quicker for lots of reasons.

One big reason is the Square Kilometre Array – which will be the biggest radio telescope ever built. All of its antennas and dishes will have a total combined area of about one square kilometre. In a decade, it will start sucking up information from the stars and galaxies. In one second, it will download more data than currently travels across the entire internet in a calendar year.

There is just one tiny problem. The data will come flooding down from the heavens at this colossal rate. But we haven't yet invented either the hardware or the software to handle this data deluge. But I am quietly confident that we will invent what we need over the next decade.

Once we do, the Snowmobile will be obsolete.

But until then, the Information Superhighway is a truck.

14

THE SMELL OF BOOKS

IF YOU LOVE reading, you'll probably have noticed that a brand new book has a rather special smell. Yes, it's that famous New Book Smell – and it's quite different from the smell of an older book. What's going on?

Book Smell comes from three sources – the paper, the ink and the glue.

Classic Old Book Smell

If you visit a traditional library with lots of really old books (pre-1850), you'll have come across the "classic" old book smell.

Researchers Cecilia Bembibre and Matija Strlič said people described the smell of these old books as "a combination of grassy notes with a tang of acids and a hint of vanilla over an underlying mustiness . . ." (Mmm, sounds a lot like any random wine description . . .)

However, a comparatively "recent" book from 1928 has been described as smelling "chocolatey, old, burnt, and of rotten socks". Why so different? Read on . . .

PAPYRUS, 4500 YEARS OLD

Books have been around for about four and a half thousand years.

Some of the earliest books come from the Fifth Dynasty of Egypt, about 2400 BC. They were written on papyrus – similar to modern paper, but thicker. Back then, papyrus was made by weaving together the stems of the papyrus plant into a sheet. This sheet was a bit bumpy, so the Egyptians smoothed out the bumps by pounding the woven sheet with some kind of mallet. Bingo – a papyrus sheet! It was generally in the form of a continuous roll (a scroll), rather than individual sheets.

Scrolls are still around – in Judaism. Jewish tradition dictates that the Torah Scroll in a synagogue must be a scroll, not a printed book. However, the congregation are allowed to use printed books.

Papyrus versus Paper

The word "papyrus" comes from the Greek word for *Cyperus papyrus* plant. And, as it sounds, the word "paper" comes from the word "papyrus".

But even though the roots of the words are very similar, the methods of production are very different.

Papyrus is made by laminating (making layers of) natural plant fibres.

However, while paper is also based on natural plant fibres, there are very important differences in its manufacture.

To make paper, the plant fibres have to be macerated (softened), or made thinner, by soaking in a liquid. They are also often pre-chopped or crushed, to increase their surface area. And lots of chemicals are involved.

BOOKS, 2000 YEARS OLD

The book, in its modern form, began to appear around the first century AD.

A book is quite different from a scroll. It's a collection of individual sheets, stuck together along one edge, which then acts as a hinge for that side. Books have many advantages over scrolls. They are easier to read, you can more quickly find a specific place, and you can stack them more efficiently than you can store scrolls.

In the early days, books were heavy and cumbersome. Each book had to be made individually by hand – with every word on every page written by a scribe. There were five different types of scribes in Europe in the Middle Ages, each with their own speciality.

So by 800 AD, a large library might have held only 500 books. (Books were very expensive.) By the end of the Middle Ages, even really large Western libraries (such as the Papal Library at Avignon, and the Library of the Sorbonne University in Paris) held only 2000 books.

Around 1045, the Chinese inventor Bi Sheng came up with movable type made of earthenware – mud or clay, not metal. Unfortunately, neither his printing press nor any books made with it have yet been found.

So (at least in the West) the credit for devising the printing press with movable type usually goes to Johannes Gutenberg. He independently invented it around 1450 – about four centuries later.

Suddenly, books could be produced much more rapidly.

Adding steam power meant that by 1800, a printing press could deliver over 1000 sheets per hour. The floodgates were now open. As of 2010, about 130 million separate titles had been published.

JOHANNES GUTENBERG

Spaces Between Words

For us today, it is obvious that in a book there will be spaces left between the individual words.

But this convention was introduced only in the 600s by Irish monks. It did not become widespread until the 1200s.

How come?

One theory is that the introduction of spaces between words was related to greater literacy. Instead of speaking the words aloud as they read, people were now better able to read silently. Because they could now read faster (because they could think faster than they could speak), they needed "assistance" to scan the text quickly – the spaces between the words.

Is this an early example of how, as a technology becomes more widely used, it can also become easier to use?

PAPER PRODUCTION

The Chinese first came up with their version of paper (similar to our modern paper) around 200 AD. Their invention slowly spread westward to Muslim countries, and then to Europe.

The book was essential to the Islamic Golden Age (mid 700s to mid 1200s AD). Many cities in the Middle East made and sold books. Yaqubi (who died around 897 AD) was a Muslim geographer and historian of world culture. He wrote that in his lifetime Baghdad had more than 100 booksellers. When paper finally arrived in the West in the 1200s, it was initially called "bagdatikos" – because it came from Baghdad.

Paper is made from wood, rag (usually linen, hemp and cotton) and grasses – not from papyrus. However, there's a big difference between the starting point (a lump of wood) and the final product – a sheet of paper. The wood has to be crushed, and lots of different chemicals have to be added, in the right quantities, and at the right stage. Some of the stages include making the wood fibres swell, removing acid, bleaching the paper to the desired grade of whiteness, conferring some degree of water resistance, adding a sheen to the surface of the paper, and so on.

Each of the chemicals added has its own odour – and they each break down, and emit gas from the paper, at different rates.

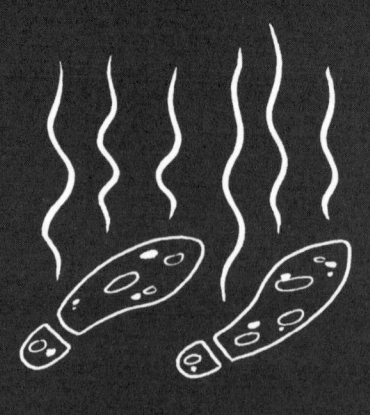

Degradic Footprinting?

Both the food and pharmaceuticals packaging industries have been pouring lots of money into studying the odours of paper and cardboard.

They're concerned that during packaging, storage and shipping, taints and odours might transfer from the paper and cardboard into their products.

This has started a whole new field of study, with its own special jargon.

This field is called "Material Degradomics", and studies all the chemicals released (known as the "Degradome") from the host product as it degrades. If you analyse the quantities of chemicals from a specific product, that's "Degradic Profiling". If you then build up a large database of how these chemicals change over time, and in response to their environment, that's "Degradic Fingerprinting". When you study what these chemicals do to the environment around your smelly product, then that's "Degradic Footprinting".

Who knew that such a field existed? Not me!

Smells for the Future

In general, our knowledge of the past excludes smells. But Japan takes a different stance on this.

In 2001, the Japanese Ministry for the Environment set up the "100 Most Fragrant List". This list came from a nationwide consultation in which local groups across Japan submitted some 5600 "candidate smells".

These aromas include sake distilleries, sea breeze, ancient woods, and (you guessed it) a street lined with bookshops. (What a cultured people!)

These 100 Most Fragrant Aromas (and their sources) are now protected. They carry a seal that translates as "scents to be handed down to our children".

A few museums around the world are now bringing in smell as part of the "museum experience". For example, the National Museum of Australia had a sensory station that "included 'the pungent smell of dried sea cucumber' alongside cooking tools of Trepang fishermen".

PAPER CHEMISTRY

Let's start by looking at the natural chemicals in paper.

There are two major sets of chemicals in woods and plants that affect the structural integrity, and smell, of paper.

The first (and minor) one is lignin. Lignin is a lot of cross-linked smaller chemicals (for example, phenolic polymers) – but in three dimensions. In other words, the smaller chemicals have been joined to each other to look a little like a chicken-wire mesh, and then laid down on top of other chicken-wire meshes. Lignin gives rigidity to the plant and doesn't rot easily.

Unfortunately, as lignin ages, it gives a yellow tinge to paper. There is more lignin in newsprint than in finer papers.

The second chemical is cellulose. It's the major component of paper. Cellulose is a long chain of lots of individual glucose molecules linked together – many hundreds, or even thousands of them. Cellulose makes up about 90 per cent of cotton, and about 45 per cent of wood.

The main chemical reactions that age paper are acid hydrolysis of cellulose and lignin, and oxidation of cellulose. These reactions have been measured to create at least 36 volatile organic compounds – of which half a dozen turn out to be markers of age.

Imagine that you have a stack of sheets of paper, say a few centimetres thick. In general, the surfaces or edges that are exposed to the atmosphere (and therefore to oxidation) are usually more degraded than the centres of the sheets (that are protected from the atmosphere). But in some cases, chemicals internal to the paper itself cause degradation. Here the centre of the page is more damaged, because the volatile organic compounds cannot escape.

The byproducts of lignin and cellulose breaking down include toluene (which gives a sweet odour), vanillin (which gives, you guessed

THE SMELL OF BOOKS

it, a vanilla odour), benzaldehyde (which smells like almonds), guaiacol (smoky aroma), and 2-ethyl hexanol (slightly floral).

Another breakdown product (furfural, which also smells like almonds) can be used to work out the age of the paper. So if you try to sell your freshly forged hand-written "100 per cent genuine" diary of the Buddha, make sure that you use paper that is older than a few years – or else your deception will be quickly detected.

The Books of Tomorrow, Today

Today's books won't degrade as quickly as older books, because they have a much lower lignin content than older books. So any breakdown products would be produced in lower concentrations. So in a century from now, today's books will smell different from how a century-old book smells today.

Thousand-Year-Old Paper

Isn't it weird that a material as fragile as paper can last for thousands of years?

Matija Strlič et al. say that unfortunately, "most paper produced between 1850 and 1990 is likely not to survive more than a century or two, due to (its) inherent acidity". This more modern paper is deteriorating rapidly – as compared to older historic papers, produced before the advent of paper mills.

One cause of this acidity is rosin. It was added to make the fibres of paper more hydrophobic (water-hating or -repelling), so that the paper was more writable. Rosin is a strange mixture of 90 per cent various acids, and 10 per cent various fats. Aluminium sulphate (which is acidic) was used in blending the rosin into the paper.

Happily, the technology of paper manufacture changed in the 1980s and 1990s. This was due to environmental concerns about the chlorinated chemicals in the manufacturing process. So (purely by a happy accident) today's paper should survive better than paper from 1850 to 1990.

But all paper is at some risk from acidity. This is because cellulose itself (in all paper) produces acetic, formic, oxalic and lactic acids as it degrades.

INK AND GLUE CHEMISTRY

There are many different inks used in books. Some inks fade with time, while others get darker. They all have their own odours.

There's also a huge range of glues that can be used to bind the sheets together along one side to form a hinge. Different glues are used to attach the book's cover, depending on whether it's hardcover or paperback.

So you can see that there is an enormous range of different chemicals that can waft out of the freshly opened pages of your brand new book.

Even within the books of one publishing house, there would be many different chemicals used, depending on the purpose of the book. A cheap black-and-white paperback would use quite different chemicals compared to an expensive coffee table book with glossy colour pages.

The whole situation gets more complicated again as the book degrades. Not only will the inks and glues age and break down at their own individual rates, so too will the lignin and the cellulose in the paper.

Today, the evolution of books has continued – away from paper, and into the digital revolution, with ebooks.

Ebooks might be convenient and light in your suitcase, but you can't give your copy to another person as a present, and they don't smell as nice.

15

EMERGENCY EMERGENCY PHONE CALLS

IN A LARGE-SCALE emergency, everybody hits the phones. Mobile phone networks can get so congested you can't get through. This congestion can be caused by damage to the network hardware, and also by increased call volumes.

But an emergency is exactly when you need voice communications the most.

WIRELESS PRIORITY SERVICE

This is why the USA has installed a "layer" over the regular mobile phone system called Wireless Priority Service, or WPS.

It has to meet the needs of the emergency response community. This means providing communications to manage national security and emergency situations – especially in the first few days following such an event.

The US Department of Homeland Security has defined five levels of priority of access to the Wireless Priority Service.

PRIORITY LEVELS

Priority 1 starts with the President of the United States and goes down through state governors to city mayors. Priority 2 covers federal and state emergency services, and the military. Priority 3 encompasses those involved in public health, safety and law enforcement. Priority 4 has the managers of public works and utility infrastructure. And Priority 5 is for those involved in longer-term disaster recovery – after the initial response has been accomplished.

WPS ACCESS

Once the user has been authorised to access the WPS system, they have to pay for it. Phone activation costs about US$10. Then there's a monthly fee of less than $5 per phone, and actual usage fee of less than $0.75 per minute.

They make their WPS phone call by dialling *272, and then the phone number.

The WPS system does not throw anybody off the network. Instead it waits for the first gap and then gives that gap to the next authorised WPS user.

For example, during the 2013 Boston Marathon bombing, the phone network immediately got so congested that most callers couldn't get through. But over 93 per cent of WPS emergency mobile phone calls got through.

Isn't it strange that approved priority service providers have to pay more to be able to provide the Emergency Service?

16

LUSCIOUS LIPS

IF YOU FOLLOW Social Media, especially of the Kardashian Kind, you'll have seen human lips of unbelievable shapes and sizes. You might even have become a little confused as to what the Ideal Lip looks like. (Yup, just one more thing to be confused about in this crazy world.)

Luckily, plastic surgeons have done a Lip Survey to let us know what's hot, and what's not. The bottom line is that the bottom lip should be twice as fat as the upper lip.

LIPS 101

In most non-mammals, the lips are fairly unimportant. They're often just a few folds of tissue hanging around the outside of the jaws. In fact, turtles and birds have given up on lips entirely – their lips have evolved into hard, solid beaks of keratin.

But in mammals, the trend has gone the other way, and lips have become very important.

MAMMALIAN MOUTH CURTAINS

Human lips are soft and highly mobile body parts – and can do so much. They have about a dozen muscles attached to them, so that they can perform complex actions.

On the practical front, lips keep food inside your mouth. They can shut so tightly that solids, liquids or gases can be kept either in or out.

We humans can wrap our lips so firmly around an uneven surface that a baby can breastfeed, and an adult can suck up a very gluggy thickshake through a drinking straw.

Besides feeding ourselves, we can create huge numbers of different sounds, such as whistling – and even play musical instruments.

Baby lips are a very sensitive tactile sensory area. They have receptors for different types of touch (fine, coarse, vibrating, etc.), heat, cold and pleasure. This is why the littlies are forever shoving things into their mouths.

Adult lips can be good for kissing. The shape and size of the lips is linked to being sexually attractive. And the lips are an erogenous zone that can be stimulated by touch.

In general, full and well-defined lips are linked to youth, health, attractiveness and sensuality. With age, there is a loss of volume, a flattening of the philtrum (between the top of the upper lip and the bottom of the nose), and some widening of the Cupid's bow on the upper lip.

Lips Are Red

On most of your face, your skin has up to 16 separate layers of cells. But the skin on your lips is much thinner – only three to five layers of cells. This thinness lets the red blood cells shine through, and gives the lips their glamorous red colouration.

Lips Dry Out

Now I'm not trying to be perverse.

But it is kind of unfortunate that the skin of the lips lacks both hair and sweat glands.

Missing out on hair and sweat glands means your lips miss out on the normal layers of body oils and sweat (which protect the skin on the rest of the body). And because they dry out more quickly, you can get chapped lips.

LUSCIOUS LIPS

For a plastic facial surgeon, there can be many goals in lip surgery – besides repair after trauma or an injury. These other goals can include restoring the volume, or fullness, of the lips to counteract the inevitable atrophy that comes with increasing age.

There are several different popular cosmetic procedures to boost the size of the lips. They include using short-term injectable fillers such as hyaluronic acid, as well as longer-term fillers such as structural fat from other parts of your body, and even soft synthetic implants.

You might be surprised to know that, even after centuries of lip surgery, the aesthetic proportions of the lips (their effect on facial attractiveness) are still poorly defined. It's all been up to what the particular surgeon thinks looks good while doing the surgery. (If surgeons had better guidelines, some "celebrities" might have better lips.)

SURVEY TIME

This is why several surgeons from the Department of Otolaryngology – Head and Neck Surgery at the University of California in Irvine carried out a survey. The goal was to assess what would be the most attractive lip dimensions of young white women. (I guess that's who their patients were.)

The surgeons started with images of 20 white women, aged 18 to 25 years. In the first part of the study, they used as judges 20 people from a conventional focus group, and then 130 people from internet-based focus groups.

In the first part of this survey, the surgeons used Photoshop-type technology to make the upper and lower lips simultaneously appear either both bigger or both smaller. They created five options for each of the 20 women's photos, giving 100 faces overall.

The 150 judges had the job of looking at the 100 faces, and then ranking them as being more or less attractive. It turned out this group of judges thought that the most attractive lips were about 50 per cent bigger than what the women actually had. This "Most Appealing Lip Size Option" filled up about 10 per cent of the lower face.

In the second part of this survey, the surgeons again manipulated the lips in the images with software. This time there were 20 judges from a conventional focus group and 408 people from internet-based focus groups.

First, they cranked up the volume of both lips so that they were about 150 per cent, or filled about 10 per cent of the lower face. Second, they varied the sizes of the upper and lower lips relative to each other. The range was that the lower lips could be half as big as the upper lip, and up to three times bigger. They picked the 15 most attractive faces, and made four versions of each face. That gave them 60 faces to be compared.

Overall, the 428 judges rated the most attractive lips to have a lower lip twice the height of the upper lip.

It's hard to get that plumped look with only a touch of lippy. So to pay full Lip Service to this survey, get out your duckface, but don't go the full Kardashian.

Limitations of This Study

One problem with this study is that it started with only 20 original faces that were digitally manipulated – a very small sample size. Furthermore, the people in the study were a very tiny subgroup of society (white females 18 to 25).

The authors chose only a few ratios of lip sizes. They may well have missed out on the most attractive, if it wasn't one of the tested options. The authors did not check if the same judges always chose the same ratio of lip size across all faces. In addition, there are many other factors to consider in trying to measure "attractiveness", such as proportion, chin position, subjectivity, etc.

But you have to do the first study before you can do the second study – so this study is a start.

17

SOLAR ENERGY PAYBACK TIME

SOLAR ENERGY PAYBACK TIME

FAKE RUMOURS ABOUT the long Energy Payback time for solar panels still keep popping up. The usual claim is that you never get back the energy that went into making the solar panels, which last only 10 years. That claim is doubly wrong.

LIFESPAN – 25 YEARS

In real life, you can count on your solar panels working for 30 years. Solar panels come with a warranty that they will operate for 25 years, and that at the end of 25 years they should still be delivering 80 per cent of their original output. That's really reassuring. Some saucepans also come with a 25-year warranty – but they are way less complicated than solar panels.

The claim that the lifespan of solar panels is only 10 years is incorrect.

THE SOLAR SAUCEPAN = TAKES 25 YEARS TO COOK SOMETHING

ENERGY PAYBACK?

There's an old saying, "You gotta spend money to make money." The new version is, "You gotta spend energy to make energy."

It takes energy to make solar panels. But, over the lifetime of the panels, you get back between 9 and 19 times as much energy as was spent to make them – assuming a 30-year life.

About 85 per cent of the energy used to make a solar panel array goes into fabricating the special silicon semiconductors. Of the rest, about 5 per cent is used to make the aluminium frames to hold the silicon, and about 10 per cent is used for the glass, the wiring, the inverter and other bits and pieces.

The main factor that determines Energy Payback time is how much sunlight you get (which is related to how little cloud cover there is in your local area).

The shortest payback time in the whole world happens in Perth – about 19 months. In other words, in Perth, the time taken for the solar panels to generate as much electricity as it took to make them is just 19 months. So that means for the next 28 years of your solar panels' lifespan, all the electricity generated is free and clean.

In Sydney and Brisbane, the Energy Payback time is about 22 months (more cloud cover). The longest Energy Payback time in the world is in Brussels (40 months). But even in this worst setting, you still get 26+ years of free energy.

SOLAR PHOTOELECTRIC 101

The Photoelectric Effect (i.e. turning light into electricity) was first discovered way back in 1839, by Edmond Becquerel. (His son Henri Becquerel also "discovered" radioactivity.) It took until 1905 for Albert Einstein to explain how light could be turned into electricity. He won a Nobel Prize for his discovery of the Law of the Photoelectric Effect.

You've probably heard of "conductors", which are very good at carrying electricity. They include metals such as copper. And you've also probably heard of "insulators", which are very bad at carrying electricity. These include glass.

It turns out that there is something halfway between a conductor and an insulator. It's called a "semiconductor". Silicon can be a semiconductor. Shining light onto a specially prepared and very thin slice of silicon can give you electricity.

Back around 1970, the first solar panels were made. They were used in space – to provide power for satellites. Between 1970 and 2015, the price of electricity from solar panels dropped from US$150 per Watt to 60 cents per Watt. And the amount of solar (photoelectric) electrical generating capacity installed on planet Earth has risen from 1000 Watts up to 300 billion Watts.

With each doubling of installed photovoltaic capacity, the technology has advanced. The increased efficiency means it takes about 12 per cent less energy to make a solar panel with each doubling.

Today solar panels generate about 1 to 1.5 per cent of the world's total electricity, and are getting cheaper and more efficient.

What's not to like?

18

POKIES, TURNING THE TABLES

AUSTRALIA HAS AROUND one quarter of all the poker machines in the Known Universe – even though we're just a tiny fraction of the world's population.

About one third of Australians "play the pokies". Each year, they lose around $10 billion. (That's a lot more than just "loose change".)

But there's an unexpected benefit to studying Maths – for pokie players.

Some Mathematical Russians worked out how to turn the tables, and win money from the pokies (instead of losing). But mostly with Australian pokies . . .

Oz Pokies

About half of Australia's pokies are in New South Wales.

The average player loses about $400 per year.

The average problem gambler loses $12,000 each year.

POKER MACHINE 101

Poker Machines were invented way back in the 1890s.

They have different names in different parts of the world. They're called "slot machines" in the USA and Canada, and "fruit machines" in England.

They were also called "one-armed bandits". The original mechanical pokies were operated by a single lever (or arm) on the side of the machine, and they took your money – hence the name.

But that's all gone. Nowadays, a typical pokie might have five columns, or reels. Today, the gambler feeds money into the poker

POKIES, TURNING THE TABLES

machine and then presses a button to activate the spinning reels. In the old days, they were mechanical reels that would physically spin, but today they are virtual reels on a computer screen. If you take the cover off a modern poker machine, it looks like the inside of a computer – because it is.

The symbols on the reels could be numbers, or fruit (such as cherries or lemons), or other symbols (such as diamonds). Depending on where the reels stop spinning, you might get a payout. For example, an unbroken set of sevens in a row might give you the maximum payout.

In Australia, the pokies have a Payout Percentage (also called the Return to Player) of at least 87 per cent (usually around 90 per cent). That means that, averaged out over a long time, the pokies will pay out, or return, 90 per cent of the money put in. So the punter gets back only 90 per cent of what they put in. But when they put that lesser sum back in again, they get 90 per cent of that. Repeat often enough, and they'll end up at zero.

The longer you play, the more likely you are to lose all your money.

Buy Some Random Numbers?

Truly random numbers are quite valuable. For example, cryptographers could use them to help keep your secrets safe.

In 1927, the Cambridge University Press published a table of 41,600 supposedly random digits.

In 1955, the RAND Corporation published A Million Random Digits, based on an electronic simulation of a roulette wheel.

143

RANDOM IS MORE THAN A WORD

Now it's important (from the perspective of the House, rather than the poor gambler and their unfortunate family) that the gambler cannot predict when a winning sequence will appear.

Ideally, the symbols should appear in a totally random order. But there are two problems with that.

First, if they were truly random, then the House would not have any advantage. It wouldn't get its cut of the gambler's money. This is why the House uses an "algorithm" (or formula) to slightly change the odds in its favour, and give itself an advantage.

Second, it's really hard to get anything truly random. Some events are indeed truly random – such as which specific radioactive atoms will decay, or the appearance of a quantum fluctuation in a vacuum. But these random events happen much too slowly and are too hard to measure for modern high-speed pokies.

PSEUDO-RANDOM

So, today's poker machines usually rely on "pseudo-random number generators". (These numbers have to be tweaked with an algorithm to give the House an advantage.)

The word "pseudo" means "not genuine" – but these pseudo-random number generators are fairly close to random.

So how does a pseudo-random number generator work?

One way is to use a simple mathematical formula based on a single number called the "seed".

For example, the seed might be a number with 20 digits.

You square that number to get a 40-digit number. Take the first 20 digits, and multiply them by the last 20 digits, to give you another 40-digit number. Repeat this process 100 times, and only then pick the middle five digits.

These then become the display on the screen the gambler sees.

Repeat. Repeat. Etc. (By the way, today's pseudo-random number generators are far more complex than this.)

PUTIN STARTS IT OFF

How did Mathematicians beat the pseudo-random number generator – and the House?

The initial driving force was Vladimir Putin. In 2009, he banned practically all gambling in Russia.

His reasoning was that organised crime in Georgia (a country immediately south of Russia) got its money from poker machines. His action was an attempt to separate the gangsters from their money supply.

As a result, thousands of Russian casinos sold their poker machines very cheaply to whoever would take them off their hands. Some of the pokies went to a team in St Petersburg in Russia. They started trying to reverse-engineer the workings inside the pokie.

They succeeded.

RUSSIAN GAMBLERS IN AMERICAN CASINOS

A few years later, small teams of Russian gamblers began taking unusual payouts on poker machines in casinos in the USA. There were a few red flags for the casino operators.

First, while a Russian gambler was playing, a poker machine would (over a few days) pay out a lot more money than it took in. That was suspicious enough! Even more unusually, these pokies did not pay out any really large sums. (The casino people call this a "negative hold" – and it's occasionally possible, but very unlikely to happen often.)

Second, the Russian player would always keep one hand in their pocket or a small satchel or purse.

Third, the Russian gambler would make lots of small plays (one cent, one cent, one cent) repeatedly, and then, seemingly out of the blue, they would make one large play of $20 or $60. And for that large play – and only that play – they would win a large amount, say $1300.

Finally, the casinos noticed the Russians were targeting only one specific machine – the Aristocrat Mark VI, manufactured in Australia. These old-ish computerised pokies had no Bluetooth or WiFi connectivity. There was an Ethernet cable going in, but it was buried behind panels. Ripping off the panels would be kind of easy to notice in a casino with all their security and video surveillance. So they ruled out that kind of interference straight away.

MATHS WINS – IN THEORY

So how were the Russian players cheating these poker machines?

They were relying on some painstaking mathematical work back in Russia.

The number sequence flashing onto the pokie screen should have been almost totally random. But the Clever Russian Maths People had worked out the exact (and therefore, not at all random) sequence of 5-digit numbers. But it was a very long sequence – millions of entries long.

So to find where they were in the sequence, they needed to first load in a few dozen "numbers" (such as 74953, 84586, etc.) as they popped up on the screen. That was enough to tell them where they were – and therefore what would come up next. Then, all they had to do was wait for the numbers that gave big payouts, such as 77777.

Cheating – Old School

Way back in the past, when the pokies were still fully mechanical and not at all electronic, the Old School cheaters worked their way through four generations of cheats.

The first two worked on the slots where the gambler inserted the money, while the last two worked at the payout chutes where the money came out.

Number One was "fast-feeding". The gambler simply shoved their coins into the slot as quickly as they could. If they did it "correctly", the pokie would give them more credit than they had put in.

"Stringing", Number Two, was slightly more advanced. They threaded a string through the coin, and simply jiggled the coin up and down past the internal reader. Each time it went past, the pokie would register that another coin had been inserted.

The poker machine manufacturers soon worked out how to stop the cheats from interfering with the money on the way in. So the cheats then moved to where the money comes out – the payout chute.

Number Three was the "monkey paw" – a flexible rod with a claw on the end of it. The cheater would slide the monkey paw up the payout chute, and wriggle the claw until it hit the lever that counted how many coins came out. Pressing that lever repeatedly got them lots of coins.

The poker machine manufacturers counteracted the monkey paw by swapping the physical lever for an optical "lever". The next step was obvious.

Enter Cheat Number Four, the "light wand". It was just a tiny light bulb on the end of the wand. Like the monkey paw, it was shoved up the payout chute in order to trick the optical sensor.

But then the manufacturers got rid of the mechanical components, and turned the pokies into computers with screens. They made pokies cheat-proof – or so they thought.

MATHS WINS – IN PRACTICE

A Russian gambler would come to an Aristocrat Mark VI – and play one cent. A set of five symbols would appear on the pokie screen.

In their pocket, their fingers would dance across the screen of their iPhone, coding symbols into an app called Severin. For example, a single tap followed by a swipe to the right might mean "lemon", a single tap followed by a downward swipe might mean "cherry", and so forth. (By the way, you can't get Severin from the Apple Store.)

Then they would put in another one-cent coin, see what appeared on the screen, and key those five symbols into Severin. They would do this a few dozen times. All this information would go back – in real time – to the criminal mothership in St Petersburg.

Back in Russia, the Mathematical Master Control would then work out where in the pseudo-random number sequence that particular Aristocrat Mark VI machine was. The gambler in the American casino would keep putting in cent after cent, until the computer in St Petersburg told them to play big.

POKIES, TURNING THE TABLES

The phone would vibrate, and a quarter of a second later the gambler would put in a big play – and get around $1000. (They were careful to avoid the major payouts, hoping to not attract too much attention to themselves.)

Cheating – Really Old School

When I was a taxi driver, I would work the night shift.

On a few occasions, I picked up the staff (not the patrons) from a specific RSL Club around 3 a.m. They would lurch out, carrying a sturdy galvanised iron bucket in each hand. The buckets were full of 20-cent coins. (Back then, you could buy a beer for 40 cents.) The back of the taxi would drop, as they loaded up the boot with buckets of heavy coins.

CASINOS DON'T LIKE TO LOSE

A team of four gamblers would typically milk the pokies of a quarter of a million dollars in a single week.

Astonishingly, it turns out that American casinos don't like to lose money. (Who'd have thought?!)

By mid-2014, some of the Russian pokie players had been picked up by the California Department of Justice. At first, the cops couldn't work out how the Russian gamblers had done it. And because it's not a crime to win in a casino and be a Russian, the cops had to let them go.

But once they worked out how the Russian players had used mathematics to win at the pokies, they started jailing them.

Today, there are still around 100,000 Aristocrat Mark VIs operating around the world. So how lucky do you feel?

It takes a one-armed bandit (a gambler with their hand in their pocket) to beat a one-armed bandit (a pokie).

Cheating – American Version

Some Americans decided to use mathematics to win lots of money as well.

In the USA, the multi-state lottery game called Hot Lotto would regularly have first prizes of over $10 million. One employee, a former Information Security Director of the Multi-State Lottery Association, allegedly rigged the pseudo-random number generator to generate very specific numbers on certain days of the year.

In 2015, he was sentenced to 10 years in jail, for his lotto ticket that he tried to cash in for $14 million. But at the time of writing, the appeals were continuing.

POKIES, TURNING THE TABLES

The Aussie Experience

We Australians make the most sophisticated and technologically advanced poker machines on the planet. Most American casinos tend to run older-style machines.

A pokie used to cost the club $28,000. A small club would pay that back in 12 months, while a bigger club could take just 6 months. Today, they're cheaper, and the payback time is about half.

If you want to play a pokie, play on a brand-new machine. It is more likely to pay out, as it has no historical data. Today's "intelligent" algorithms analyse cash-in versus cash-out in real time. So, if there's a big payout, the Payout Percentage drops from 90 per cent to (say) 88 per cent – for a while.

In Oz, each reel has 24 symbols. If there's a big payout, the number of symbols can increase, to reduce the chances of winning.

Companies in Melbourne sell "really nice" pseudo-random number generators, and the algorithms for the payouts – and get paid up to a million dollars.

Social Cost of Poker Machines

Back in the days of coin-operated poker machines, there was a limit to how much you could lose in a single session. But with the modern (and far more addictive) electronic poker machines, that have "limits" per bet of up to $10, you can lose much more than just small change.

Australia has about one quarter of all the poker machines in the world. About half of those are in NSW. The tax on poker machine profits gave the NSW government over $1.5 billion in 2016-17 – about 5 per cent of its total taxation revenue. A tiny fraction of this revenue is earmarked for a Statewide Infrastructure Fund – to be spent on sporting, community or health infrastructure. However, the "party line" is that gambling revenue is reinvested in local communities.

The gambling industry in NSW is very predatory – and Fairfield is ground zero.

Fairfield has about 2.6 per cent of the NSW population, but provides 7.9 per cent of poker machine profits for NSW.

Fairfield is the most disadvantaged area in Sydney – by the markers of education, average income, family stability and English-language skills. The gamblers of Fairfield have poured increasing sums of money into their poker machines – $6.7 billion in 2014-15 and $8.27 billion in 2015-16. The average Fairfield resident has annual gambling losses of $4,171 – greater than 10 per cent of the annual average local income of $39,936.

And yet, none of the tax revenue taken from Fairfield has been spent (via the Statewide Infrastructure Fund) on facilities in that council.

19

GOLD NOBEL MEDAL INVISIBLE TO NAZIS

IS THERE ANY possible downside to winning a Nobel Prize? Well, yes, it can be very dangerous.

HAPPY ENDING

A few years ago, a colleague of mine was heading home from visiting his grandmother in the town of Fargo in North Dakota, USA. He was at the airport, and his carry-on luggage was passing through the X-ray scanner.

This colleague, Professor Brian Schmidt, had just won a Nobel Prize.

The Nobel Prize medal is made of gold, which is very dense – indeed, one of the densest metals. The X-ray scanner at the airport was showing a completely black circle inside his computer bag – which is very unusual. The gold (19 times denser than water) was blocking the X-rays completely.

The Transport Security Administration (or TSA) officer said, "There is something in your bag."

Brian Schmidt opened his computer bag and pointed at the little carry case that contained the Nobel Prize medal and said, "I think it's probably this."

The TSA officer said, "What is this?"

And Brian replied, "A large gold medallion."

Now like me, and most people on Earth, the TSA officer had never actually seen a gold Nobel Prize medal. So he asked, "So, what is it made of?"

"Gold."

"Who gave this to you?"

"King Gustav of Sweden," was the reply.

The TSA officer asked, "Why did he give it to you?"

And Brian replied, "Because I was the leader of a team that

discovered that the expansion rate of the universe was accelerating."

The security officer replied, "It's a Nobel Prize? You're a Nobel Prize winner?"

And Professor Schmidt replied, "Yes."

Once the TSA officer accepted that Professor Schmidt was a genuine real-life Nobel Prize winner, and worked out that he had just carried his gold Nobel Prize medallion across the world to proudly show it to his grandmother, everything was fine.

TROUBLE WITH GOLD MEDALS

But on one occasion, having a gold Nobel Prize medallion in your luggage with the winner's name on it could have been a death sentence.

This was the case in April 1940, when the Nazis invaded Copenhagen.

Some years earlier, two German Nobel Prize winners (Max von Laue, a Jewish sympathiser and opponent of the Nazis, and James Franck, of Jewish heritage) had illegally sent their gold Nobel Prize medals from Hitler's Germany to the legendary physicist Niels Bohr in Copenhagen for safekeeping. Sending gold out of Germany was a crime. But they did it anyway, so that the Nazis couldn't get their hands on their gold Nobel medals.

However, once the Nazis invaded Copenhagen, the medals were a liability. One member of Bohr's team, George de Hevesy, realised the potential danger.

He later wrote, "In Hitler's empire it was almost a capital offence to send gold out of the country."

If the invading Nazis found the gold Nobel Prize medals engraved with the names of the German scientists Max von Laue and James Franck, that simple discovery could have led to a lot of people dying.

George de Hevesy suggested burying the medals in the grounds of Bohr's laboratory – the Institute for Theoretical Physics. But, as Niels Bohr and all forensic scientists knew, the only thing you can never hide is a hole in the ground. (A hole that has been filled in always shows a boundary between the disturbed and undisturbed soil – even after tens of thousands of years.)

TEMPORARY SOLUTION

So de Hevesy decided to dissolve the gold. He later wrote, "While the invading forces marched in the streets of Copenhagen, I was busy dissolving Laue's and also James Franck's medals."

Like all precious metals, gold is pretty inert – it doesn't mix, tarnish or dissolve. You can leave it buried in the ground for millennia, and it will be untouched by time, or by its long incarceration in the dirt. But not even gold will stand up to aqua regia, which literally means "royal water" or "king's water".

So in Denmark in April 1940, as the Nazi invaders marched through Copenhagen, George de Hevesy began dissolving the two gold Nobel Prize medals that had been illegally taken out of Germany.

And it would really have been a race against time – these medals are quite substantial. Before 1980, they were 66 millimetres across, weighed about 200 grams, and were made of 23-carat gold. But luckily, the two medals disappeared into the liquid in time. The aqua regia very slowly changed from faintly peach to bright orange.

So George de Hevesy had a suspiciously heavy beaker of orange-coloured aqua regia containing the secretly dissolved gold Nobel Prize medals. He placed it on a high laboratory shelf, well above eye level, and left it there with other bottles of chemicals. Three years later, in 1943, Nazi-controlled Copenhagen was no longer safe for a Jewish scientist, so de Hevesy left for Sweden.

GOLD NOBEL MEDAL INVISIBLE TO NAZIS

According to Sam Kean, author of *The Disappearing Spoon*, "When the Nazis ransacked Bohr's Institute, they scoured the building for loot or evidence of wrong-doing, but left the beaker of orange aqua regia untouched". De Hevesy returned to the laboratory after the Nazi defeat. His beaker of orange-coloured liquid was as he had left it – in open view.

Using some easy chemistry, he got the dissolved gold back as a precipitate. Around January 1950, he sent the raw metal back to the Swedish Academy in Stockholm, who had originally awarded these medals.

This gold was then recast in Nobel medallions, and presented anew to Max von Laue and James Franck in 1952.

The hero of our story – George de Hevesy – went on to win a Nobel Prize himself in 1943 (a lovely feel-good ending).

But it wasn't just his genius that outsmarted the Nazis.

What he needed was some basic (or in this case, very acidic) chemistry . . .

GEORGE DE HEVESY

Aqua Regia

Gold is incredibly inert. (I wrote a story on the origin of gold, and how it can end up in trees, in my 36th book, *House of Karls*.) But in the 8th century, Jabir Ibn Al-Hayyan invented a liquid that could dissolve gold – aqua regia. (While it can also dissolve platinum, this yellow-orange fuming liquid can't dissolve all metals.)

The component acids of aqua regia react with each other. So it is usually made up just before use. Otherwise, it quickly decomposes and loses its effectiveness.

Aqua regia is a 3-to-1 mixture of hydrochloric acid (HCl) and nitric acid (HNO_3). Individually, these acids can't dissolve gold (Au). But combined in this very specific ratio, they can work together (like a super-destructive tag team in wrestling). They each do a tiny amount of work – and then hand back to the other acid.

The nitric acid gets in first to attack the gold. It rips three electrons off from each of just a few gold atoms at the surface. It turns these gold atoms into positively charged gold ions.

$$Au \rightarrow Au^{3+}$$

Then this reaction comes to a complete halt. (Actually, the reaction comes to an equilibrium – a state of balance. Each instant, equal numbers of gold atoms are leaving the solid gold and going into solution – and vice versa. But overall, there is a fixed [and very small] number of gold ions floating in the solution.)

Now hydrochloric acid steps up for its turn. Unlike nitric acid, it can't attack gold. But it does provide negative chloride ions (the Cl⁻ in HCl) to the solution. Four of these chloride ions then react with, and mop up, a single charged gold ion.

$$Au^{3+} + 4\ Cl^- \rightarrow AuCl_4^-$$

So all of the first batch of positively charged gold ions are mopped up, and combined with chloride ions.

The nitric acid is now able to attack the surface of the gold again, and make another very small number of positively charged gold ions. Then the hydrochloric acid swings into action.

Repeat indefinitely, and you can, albeit very slowly, dissolve gold.

20

SPACE JUNK

W**E HUMANS ARE** a messy lot. We leave junk behind us, everywhere. Space is no different. We have already paid a price for littering in space, with damaged spacecraft.

But it can get a lot worse.

Unless we clean up our act, we may find that clouds of orbiting space junk, travelling at hypersonic speeds, could seriously interfere with our future space operations.

ISS THREATENED

Space junk has closely threatened the International Space Station (ISS).

On 16 occasions, the ISS has actually had to fire its rockets to shift its 420-tonne bulk. This was to get it out of the way of some incoming space junk that could have destroyed it.

On four occasions, the threat from space junk was not realised until there was literally no time to move the ISS out of the way. On these occasions, the crew had to put on their space suits, seek shelter in the Soyuz re-entry spacecraft – and hope that nothing blew up. If the International Space Station had been hit, they would have abandoned it, and carried out an emergency landing on Earth in the Soyuz.

The Space Shuttle used to re-supply the ISS. A Space Shuttle window had to be replaced after every second flight! Microscopic space junk kept hitting those windows.

Even worse, fully functional (and expensive) spacecraft have been destroyed by collisions with space junk.

SPACE JUNK HISTORY

Space junk is all the dead and non-functioning artificial objects that we humans have put into orbit around the Earth. The amount of junk up there in orbit is always increasing. Whenever there is a collision between a big bit and a little bit, you get another couple of hundred little bits.

Space junk includes old satellites, spent rocket stages that were used to launch another rocket or a satellite, flecks of paint, and dust from solid rocket motors.

The oldest piece of space junk is the Vanguard 1 satellite, dating back to 1958.

Also orbiting up there are obsolete nuclear-powered Russian Radar Ocean Reconnaissance satellites from the 1970s and 1980s. They were powered by BES-5 nuclear power packs. These satellites are leaking frozen sodium–potassium coolant.

Astronauts drop stuff too. Michael Collins lost a camera on the Gemini 10 mission, and the astronaut Ed White even lost a glove.

Gravity – The Kessler Syndrome

In the movie *Gravity*, a Russian rocket hits a defunct satellite. This accidentally sets off a runaway chain reaction (the Kessler Syndrome), which exponentially increases the amount of debris in orbit. This cloud of space debris destroys the Hubble Space Telescope and the Space Shuttle. This leaves just two astronauts, played by Sandra Bullock and George Clooney, the only survivors in orbit.

The Kessler Syndrome is real. But what we saw in *Gravity* was an unrealistically fast version – not the slow version that is happening right now.

Donald Kessler, former NASA astrophysicist, first proposed in 1978 that a runaway cascade of collision of space junk was possible. In 1991, Kessler published "Collisional Cascading: The Limits of Population Growth in Low Earth Orbit" in *Advances in Space Research*. He identified three scenarios.

If the density of space junk is low enough, it decays in orbit and falls back to Earth faster than new space junk is created. No worries.

At the critical density, additional space junk leads to additional collisions. The amount of space junk increases slowly. Minor worry.

The third scenario looks at densities above this critical mass. The production of space debris vastly exceeds its decay, which leads to a cascading chain reaction of collisions. This reduces the size of the orbiting population of space junk (to several centimetres in diameter), which increases the hazard. This is the chain reaction known as the Kessler Syndrome.

It seems that the Kessler Syndrome has already begun (very slowly) in the dense and critical 900- to 1000-kilometre orbit. One major worry is the defunct Envisat satellite – 8.2 tonnes, up at 785 kilometres. Every year, two known catalogued objects pass within 200 metres of it. Envisat is expected to remain up there for the next 150 years. Just one major collision with Envisat could bring us much closer to a runaway Kessler Syndrome.

Global Warming Prolongs Life of Space Junk

Global Warming has had an unexpected side effect – the time that space junk spends in orbit has now been increased!

Carbon dioxide is warming up the lower atmosphere. (This is not controversial – it's a fact.) The heat from ground level is rising, hitting the carbon dioxide, and being scattered back down into the lower atmosphere. The heat is not getting into the upper atmosphere, which is why the upper atmosphere is cooling.

As a result of Global Warming, in the last 10 years the Earth's outer atmosphere (above 100 kilometres) has become less dense by 3 per cent.

This means there's less drag on space junk – and that means that space junk stays in orbit for longer.

HOW MUCH JUNK?

There is a lot of junk up there orbiting our planet.

We estimate there are about 5000 tonnes of space junk orbiting the Earth.

After all, we have had more than 5000 space launches since the Dawn of the Space Age in 1957. A launch often delivers several objects into orbit, so these 5000 launches have given us over 30,000 large space objects.

In 2016, the best estimates were that there were billions of pieces of space junk smaller than 1 millimetre, hundreds of millions of objects between 1 and 10 millimetres in size, and about half a million objects in the range 1 to 10 centimetres. Included in this last category are about 100,000 bits of space junk bigger than 5 centimetres – many of them small enough to be difficult to detect, but big enough to cause lots of damage.

In 2017, the US Space Command's Joint Space Operations Center was regularly tracking more than 22,000 objects orbiting the Earth. But fewer than 5 per cent of these were functioning spacecraft – over 95 per cent were junk. The Joint Space Operations Center tracks the junk so they can manoeuvre active spacecraft out of the way of an impact.

USEFUL ORBITS

Various orbits are useful for commercial, military and scientific programs. Unfortunately, space junk inhabits all of them.

For example, the Geostationary Orbit is about 36,000 kilometres up. It's called "Geostationary" because satellites in this orbit take 24 hours to orbit the Earth. By a nice coincidence, 24 hours is how long the Earth takes to spin on its own axis. So an object in Geostationary Orbit will appear to hover over a fixed position on the ground, even though it is actually travelling at some 3 kilometres per second.

We have many communications, TV and weather satellites in the Geostationary Orbit.

Weather satellites in Geostationary Orbit can stare unblinkingly at one part of the Earth for weeks, years and even decades. This gives us a good long-term record, and helps improve our ability to predict weather.

A communications or TV satellite can stay in the same spot relative to the ground. So a TV viewer on the ground can point their dish exactly to a specific point in space, and know that they will pick up their favourite programs reliably.

Another extremely useful orbit is Low Earth Orbit. Objects in Low Earth Orbit circle the Earth some 15 times each day.

Low Earth Orbit ranges from 200 to 2000 kilometres above the surface. The Hubble Space Telescope is up there at around 540 kilometres of altitude, while the International Space Station (ISS) is at around 350 to 400 kilometres.

Along with this useful stuff, there are about 1900 tonnes of space junk in Low Earth Orbit. About 98 per cent of this 1900 tonnes is accounted for by just 1500 objects, each over 100 kilograms.

The orbit between 900 and 1000 kilometres is very useful – and, unfortunately, especially congested with space junk.

Depending on their orbit, bits of space junk can zip past each other, or smash into each other, at speeds of up to 17 kilometres per second – a mighty trash crash.

Destroy Satellite from Ship

It's possible to destroy a satellite in orbit, but it's not a trivial operation.

In December 2006, the US National Reconnaissance Office launched USA-193, a classified radar-imaging satellite. It reached orbit, but lost contact with the ground within a few hours. It was carrying about 450 kilograms of hydrazine (a rocket fuel) – which would have been toxic to life if it had survived re-entry back to the ground. Eventually the decision was made to blow up USA-193 and dilute the toxic hydrazine over a large area. Unfortunately, USA-193 was a difficult target because it did not have the "normal" heat profile of a warhead, which all the interceptor rockets had been "trained" to find. Even worse, USA-193 was tumbling erratically end-over-end, making it an even harder target to lock on to.

The job of destroying USA-193 was given to the powerful fighting Aegis ship USS *Lake Erie* in December 2007. It normally carried the SM-3 Interceptor rocket – about 6.5 metres long and 34 centimetres wide, and weighing 1.5 tonnes. But the SM-3 was designed to intercept short- and intermediate-range ballistic missiles – not for anti-satellite operations. It needed major changes. Within three weeks (an amazingly short time), three SM-3s had been modified to deal with the difficult target, and installed on the

USS *Lake Erie*.

On 20 February 2008, one SM-3 was launched from the Pacific Ocean. Shortly before USA-193 re-entered the atmosphere, the SM-3 smashed into it at an altitude of 247 kilometres and a closing velocity of about 36,667 kilometres per hour. The remaining two SM-3s were later reconfigured back to their original configuration. The total cost was about US$50 million.

The official reason given for the exercise was to "reduce the danger to human beings" by controlling the debris field. However, other sources strongly suggest that this was a military exercise, to test potential anti-satellite capabilities.

Legals?

A new-ish functioning satellite can cost US$100 million.

So who is at fault when an old lump of space junk destroys this shiny new satellite?

After all, neither object had the ability to manoeuvre in order to avoid a collision. The space junk had been orbiting for many years before the new satellite was launched.

EXPLOSIONS AND COLLISIONS
→ SPACE JUNK

One piece of space junk can become many pieces of space junk.

Sometimes, stuff in orbit can blow up by itself – a so-called Single Vehicle Event.

In 2006 alone, there were eight spontaneous breakups of space hardware.

In February 2015, the US Air Force Defense Meteorological Satellite Program Flight 13 exploded in orbit, creating more than 149 pieces of space junk. On 27 July 2016, the Chinese Long March 7 booster exploded, creating a fireball that was visible across half of the USA – from Utah to California.

But there can also be a major collision between large objects – sometimes deliberately.

Both the USA and the Soviet Union carried out multiple series of anti-satellite weapons tests in the 1960s and 1970s. The idea was simple. A rocket was launched from the ground, or from a high-flying jet, to deliberately destroy an orbiting spacecraft.

In 1985, a US test destroyed their own 1-tonne satellite at 525 kilometres of altitude, creating thousands of pieces of space junk longer than 1 centimetre.

In 2007, the Chinese tested their anti-satellite weapons systems by destroying one of their own weather satellites – the Fengyun 1C. This was the largest single space debris event in history. It created more than 2300 pieces bigger than 3 centimetres across, 35,000 pieces bigger than 1 centimetre across, and 1 million pieces bigger than 1 millimetre across.

Mostly however, the collisions happen by accident. On 10 February 2009, a defunct Russian Kosmos 2251 spacecraft weighing 800 kilograms smashed into the functioning Iridium 33 satellite weighing 689 kilograms. This happened at 789 kilometres above the Taymyr Peninsula in Siberia at a closing speed of 11.7 kilometres per second – over 42,000 kilometres per hour.

So big stuff can run into big stuff – creating lots of smaller debris.

But on a much smaller scale, in 2016 the British astronaut Tim Peake photographed a 7-millimetre crack in the window of the Cupola module of the International Space Station. It was caused by running into a hypersonic microscopic piece of space debris.

BURN IT UP: PARTS 1 AND 2

We need to clean up our act, and our space junk.

There are three major pathways to removing space junk, each coming from a different part of the world.

European proposals deal with the big stuff in a big way – using the space equivalent of a tugboat. Think of a giant space tug in orbit. It ambles up to, and then grabs, a large lump of space junk. The space tug then sends that space junk into a downward trajectory, so it burns up in the atmosphere.

An American proposal involves a solid shield that would mop up space junk by actual physical contact. The impact platform would be huge – the size of a football field. And it would work like a giant windscreen, with space junk (not insects) smashing into it.

Aerodynamic Drag

Aerodynamic Drag (or wind resistance) is what cars and planes fight when they push through the air. It's why cars are now shaped very carefully, to improve both fuel economy and performance.

But even a few hundred kilometres up, in Low Earth Orbit, there is still some air present. There's very little of it – about one tenth of one trillionth of what's present at ground level. But the International Space Station is pretty big (about 100 metres by 70 metres by 20 metres, weighing close to 420 tonnes) and is moving pretty fast (around 8 kilometres per second, or 28,000 kilometres per hour).

The ISS orbits at an altitude of 370 kilometres. It loses about 70 metres of altitude every day, due to Aerodynamic Drag.

On Bastille Day (aka 14 July) in 2000, the Sun had a hissy fit – an extremely powerful solar flare. It was so bright that the Voyager 1 and 2 spacecrafts, way past Pluto, could see it. Fifteen minutes after the Sun erupted, energetic protons hit the Earth. They heated up the Earth's atmosphere, which expanded. This increased the drag on the ISS. In just one day, the International Space Station lost 15,000 metres of altitude – that's 15 kilometres!

The ISS lifting rockets were put into action very quickly to recover the lost altitude.

BURN IT UP: PART 3

This approach comes from several teams in Japan and in Australia. It involves creating drag to slow down the orbiting space junk, so it falls out of orbit and burns up in the atmosphere. This is based on a new concept of "Electrodynamic Drag", not the more familiar "Aerodynamic Drag".

Electrodynamic Drag involves reeling out a long conducting string (or tether) from the space junk – some several hundred metres long. As this tether slices through the Earth's magnetic field at 28,000 kilometres per hour, it creates Electromagnetic Drag. Electromagnetic Drag could slow down space junk – and send it to a fiery end in our atmosphere.

It's a great idea – but getting those conducting tethers onto big chunks of space junk is a challenge.

One suggestion is that all new spacecraft should have a tether built into them. At the end of their life, they would reel out the tether, slow down, and burn up in the atmosphere. For pre-existing pieces of space junk, you'd somehow have to attach an "after market" tether to them, maybe with a "space tug". The aim of the tether would still be to slow the space junk down then make them fall and burn up.

Another "drag" concept involves a ground-based laser to vaporise the front (or leading edge) of the space junk. The aim is to produce a rocket-like forward thrust that would slow the space junk enough to burn up in the atmosphere. NASA research in 2011 showed that firing a laser beam at small space junk for a few hours per day could drop its orbit by 200 metres each day. However, an unwanted possibility is that the space junk could instead break up into more pieces of smaller space junk.

TRAGEDY OF THE COMMONS

The problem of space junk is a classic case of the Tragedy of the Commons.

This is an Economic Concept involving a shared common resource – in this case, Space. Space users do not act for the common long-term good of their society. Instead, each selfishly acts entirely in their own self-interest. The tragic result is that the Commons get degraded, so that future generations miss out.

The current situation is that millions of pieces of space junk, moving at hypersonic speed, could seriously interfere with future space plans.

Overall, space technology generates some US$160 billion each year. This includes revenue from military surveillance, climate and weather data, international phone calls and television broadcasts. But there is no commercial incentive to keep these useful orbits around the Earth clean, because the clean-up costs are not assigned to the polluters. And now there are plans to launch over 700 small satellites into Low Earth Orbit over the next few years.

Where's Dr Evil and the Frickin' Sharks with Laser Beams when you need them most?

Don't Know Where Satellites Are

Surprisingly, the exact orbits of many satellites are not generally known. (And when satellites turn into space junk, that hard data becomes even less well known.)

Why? Commercial and strategic grounds.

If your opponent/competition knows the exact positioning of your spacecraft, that information can reveal the extent of your capabilities. Unfortunately, this deliberate process of **not** revealing the exact orbit introduces fuzziness into working out collision paths. But having collisions is (obviously) bad.

There are a few solutions.

One is to give exact orbital information to a trusted third party. The Major Problem is finding that trusted third party.

Another, and more achievable, solution is to use Mathematics. (See, here's yet another use for Maths, besides credit card security and videos of cats.)

This solution comes from DARPA. That's the USA's Defense Advanced Research Projects Agency – you know, the folk who gave us the internet, Graphic User Interfaces for computers, driverless cars, etc. DARPA has developed Secure Multiparty Computation Protocols to solve this problem. They involve lots of encryption.

Very cleverly, these DARPA Protocols allow the various operators of the satellites to share their encrypted data (the exact orbits) onto a shared computer. This orbital data is then analysed to give a desired outcome – the possibility of a collision. And yet, none of the other parties get access to anybody else's exact orbital data.

However, there is a disadvantage. The intense mathematical cryptography required slows down the computations to estimate the likelihood of a collision, from milliseconds to 90 seconds. This longer time would limit the number of potential collisions that could be anticipated. However, this is purely a theoretical problem at the moment – nobody wants to use the DARPA Protocols.

Whipple Shields

Whipple Shields are a great example of that famous Engineering saying, "Don't re-invent the wheel". In this case, a lesson from Nature protects the International Space Station from tiny space junk.

The story begins in 1946, when the Atmospheric Scientist Helmut Landsberg collected small magnetic particles at ground level that were associated with the Giacobinid meteor shower. This intrigued Fred Whipple, an American astronomer. Fred then showed that as these particles hit the atmosphere, they very

THE FIRST PROTOTYPE WHIPPLE SHIELD

quickly slowed down and became harmless – like tiny dust. They fell to Earth un-melted, due to their tiny size and low velocity. Fred Whipple then invented the phrase "micro-meteorite" to describe them.

From observing micro-meteorites, he realised he needed to work out how to slow down small space debris – to make it harmless. He then developed the concept of a "meteor bumper", now called the Whipple Shield.

The Whipple Shield is a thin film of foil, held a short distance away from the spacecraft. When a micro-meteoroid hits the foil, it vaporises into a plasma that quickly expands and dissipates. By the time this plasma reaches the actual skin of the spacecraft (a few centimetres away), it has become so spread out that it does not penetrate the skin. (But if the micro-meteorite had remained as a point-like rock, it would have had easily enough energy to punch through the wall of the spacecraft.)

Later research has shown that with regard to particles travelling at hyper velocity (about 7 kilometres per second), ceramic-fibre woven shields offer better protection than aluminium shields of the same weight.

Another version uses gas-filled chambers. This has the advantage that upon impact, the pressure wave spreads in all directions – both directly onto the spacecraft as well as laterally and backwards. This reduces the impact delivered to the spacecraft.

So today, the ISS is shielded to withstand impacts with objects up to 1 centimetre in diameter.

Space Junk Hits Ground

Most incoming space debris burns up in the atmosphere. But a surprising amount makes it down to the ground.

Over the last half-century, an average of one catalogued piece of debris has fallen back to Earth each day.

But this varies with the activity of the Sun. The Sun gets hotter and colder – by 0.7 per cent – over an 11-year cycle. (No, this is not the cause of Global Warming.) The hottest time is called Solar Maximum, while the coolest – about 5.5 years later – is called Solar Minimum. When the Sun is at its hottest, the Earth's atmosphere expands, and gives more Aerodynamic Drag to objects in orbit.

This rate of space-junk-from-Heaven varies from almost three objects per day at Solar Maximum, down to one object every three days at Solar Minimum.

In 1969, five sailors on a Japanese ship were injured by space debris when it crashed into their ship.

In 1997, Lottie Williams in Oklahoma was luckily not injured when she was hit on the shoulder by a 10-centimetre by 13-centimetre piece of blackened, woven material. This was part of the propellant tank of a Delta II rocket, which had been launched in 1996.

In 2003, the Space Shuttle Columbia disintegrated on re-entry. Large parts of the spacecraft, and entire equipment systems, reached the ground intact.

On 27 March 2007, the pilot of a LAN A340 flying between Santiago and Auckland saw debris from a Russian spy satellite pass by him within a kilometre – amazingly close. He also heard a sonic boom.

Tracking Space Junk

The first part of dealing with the ever-increasing cloud of space junk is to know where it is. The technical problem to overcome is that it can be small, fast-moving, and very far away.

From 1995 to 2002, the NASA Orbital Debris Observatory (in New Mexico) tracked space junk using two telescopes. One had a 3-metre liquid mirror. In other words, it had a 3-metre wide dish of liquid mercury that was spun at 10 revolutions per minute. This pushed the mercury into a parabolic shape, perfect for a telescope mirror. (I wrote about this is in my 8th book, *Latest Great Moments in Science*, in the story "New Telescopes".)

But today, radar and optical detectors (LIDAR) are the main tools for tracking space debris. Under the best conditions, debris as small as 1 centimetre can be tracked. However, working out the orbit to allow re-acquisition of that same bit of space debris is difficult.

The US Strategic Command uses both ground-based radar and telescopes, and a space-based telescope. The space-based telescope is the Ball Aerospace Space-Based Space Surveillance (SBSS) satellite. It has a two-axis, stabilised visible-light sensor. It examines every satellite in geosynchronous orbit once every 24 hours.

Other data comes from the European Space Agency (ESA), the Space Debris Telescope, the Tracking and Imaging Radar (TIRA) system, the Goldstone, Haystack and EISCAT radars, and the Cobra Dane phased array radar.

In the USA, LeoLabs has set up a dedicated radar array in Texas (15 by 50 metres) to track space junk. This radar, combined with data from existing scientific radar in Alaska, has increased their coverage of stuff in Low Earth Orbit from 80 per cent to 95 per cent. As of March 2017, this system can track objects down to 10 centimetres. The goal is to track the quarter of a million objects as small as 2 centimetres across that are not tracked today. These 2 to 10 centimetre objects are a great risk to satellites.

21

CHILDHOOD AMNESIA

PARENTING IS ALL about nurturing. You give, give, give on so many levels – physical, intellectual, emotional and financial. And luckily, it's a selfless kind of giving – because your beloved child will remember hardly any of the first half-dozen years.

It turns out that most regular children can barely remember anything from their first six or so years of life. Welcome to the concept of Childhood Amnesia (also called Infantile Amnesia over a century ago, by Sigmund Freud).

MEMORY 101

We still don't fully understand how we remember any experiences at all.

We know that there are many types of memory. There's the memory involved in riding a bike (called "automatic procedural memory"), and the memory involved in remembering a phone number so that you can immediately dial it (called "working memory").

We do know that it's possible to implant a false memory into about one quarter of people. (I wrote about this in my 30th book, *Curious and Curiouser*, in the story "Repressed Memory".)

We also know that, unless you constantly refresh memories, they fade. The 19th-century psychologist Hermann Ebbinghaus tested human memory with "nonsense syllables" – made-up sounds such as "slan" and "kag". He found that if people learnt them, but then didn't try to remember them, half of the nonsense syllables vanished within an hour. After a month, only 2 to 3 per cent of the syllables were retained. (I wrote about the importance of refreshing memories in my 40th book, *The Doctor*, in the story "Perpetual Past".)

Childhood Amnesia is real, but like most things to do with memory, we don't fully understand it.

Naturally, there's a bunch of theories to try to explain it.

EARLY HYPOTHESES

The first theory relates to the concept of "self". Apparently, having a more robust concept of who you are helps you remember things better. For example, consider babies. If they can recognise themselves in a mirror, they are more likely to remember things such as how to play with their toy.

Another theory relates to language. A study was carried out in the Emergency Department of a kids hospital. The kids were all about 26 months old. They were all followed up 5 years later. If they could already speak pretty well when they spent their time in the Emergency

Department, they were much more likely to remember what happened to them – as compared to kids who hardly spoke at all.

The explanation for this is that if you have "language", you can both "look" and "label". This hypothesis proposes that if you can name objects, you can better remember them. So instead of being just a blur, the Emergency Department waiting room had some big red chairs, a pit full of toys, a water dispenser, and later the doctor looked at your leg with some very bright lights.

Yet another theory to try to explain Childhood Amnesia looks at the fact that when you were a child, you were also very small. So the world looked very different.

You couldn't see the top of the table, only the underneath. You couldn't see what was on the mysterious "upper shelves". But once you got past about six years of age, you began to enter the Land of the Tall People. You began to relate to the world in a very different way from when you were smaller. So your memories from back then simply didn't fit – hence the Childhood Amnesia.

LATEST HYPOTHESIS

The latest theory on what causes Childhood Amnesia relates to neurogenesis, which is the laying down of new nerve cells – such as the ones that store our memories. There's a huge amount of neurogenesis in our brains in our first half-dozen years of life. Babies can lay down 700 new nerve connections each second.

Basically, this theory says that when you lay down lots of new nerve cells, you write over, or wipe out, or recycle the old nerve cells. And when they go, so do the memories associated with them.

In other words, there is a price to being able to develop your brain so rapidly in the first half-dozen or so years of your life. The price is that, as an adult, you cannot remember what happened in those years of your life.

There is a balance between Plasticity (the ability to incorporate new information into your brain) and Stability (the ability to keep old information in your brain).

The neurogenesis research was carried out on (you guessed it) mice.

Young mice (17 days old) got an electric shock in a specific location. After a few weeks, the young mice seemed to have forgotten about the incident, and happily hung out in the place where they had been shocked. But when the scientists deliberately stopped neurogenesis with a specific drug, the young mice didn't forget, and wouldn't go back to where they were shocked.

The scientists also did the reverse study with older mice (60 days old). They gave the older mice an electric shock and, sure enough, the mice avoided the place where they had been shocked. But when the researchers *increased* neurogenesis in these older mice, they forgot the electric shock.

Some psychologists are a little skeptical as to this latest hypothesis. But however poorly we understand it, Childhood Amnesia really happens.

It's all very strange. Babies and children are sponges for new information. The first few years of our lives are very powerful in shaping who we become. And yet they are so easily forgotten.

So parents, take it on the chin, and don't get shocked (or even surprised) when your kids can't remember the blood, sweat and tears that you ever-so-lovingly poured into them.

When Does Childhood Amnesia End?

On average, as adults, we tend to remember very little of what happened before the age of six or seven. There are some patchy fragments of memory from about three and a half years of age, but very little.

But for Maori children, brought up in the traditional way, this change happens around the age of two and a half years. The Maori parents have a very elaborative style of re-telling family stories. This reminiscing not only has the social function of sharing experiences with others, it also makes the narrative more robust in the minds of the listeners – the content, the chronology and the emotional significance.

22

NATURAL ALARM CLOCK

HOW COME SOME people can accurately wake up at the same time every work day, without ever using an alarm clock? After all, one aspect of that strange state of consciousness we call "sleep" is that our normal conscious self is not paying attention, and has no control.

The answer seems to be a rapid rate of increase in one specific hormone. Strangely, some of us can "subconsciously" control this rapid increase.

THE STRESS OF WAKING

In the human body, many different endocrine glands produce many different hormones.

An endocrine organ in the brain called the Anterior Pituitary Gland is specifically related to having a natural alarm clock. This gland makes half a dozen very different hormones. One of them, rather confusingly, has several names. These names include "adrenocorticotropin", "corticotropin", and "adrenocorticotropic hormone" or "ACTH". Let's just call it ACTH.

ACTH is often produced in response to some kind of biological stress – which includes the "terrible" stress of waking up from a restful sleep. It's also strongly involved in the circadian rhythm (or biorhythm) of many organisms. It has other jobs too, such as making new bone.

ACTH has a half-life in the blood of about 10 minutes. In other words, after 10 minutes half of the ACTH has been removed from the blood. It's a protein made up of 39 amino acids.

One study looked at blood levels of ACTH in 15 healthy young volunteers. The lights were turned off at midnight, and then the volunteers were told they would be woken at either 6 a.m. or 9 a.m. In each case, just before either 6 a.m. or 9 a.m., there was a sudden and very sharp increase in the blood level of ACTH.

Now this is very surprising. Being told a number (6 a.m. or 9 a.m.) changed the timing of ACTH release!

We tend to think that "anticipation" is something that we humans do only when we are awake. But it seems that anticipation is something that can pervade our sleep.

So if coffee isn't your morning wake-up drug of choice, try ACTH . . .

23

SINKHOLES

THEY CAN EASILY swallow up your car or your house. One tenth of the Earth's land area is susceptible to them.

What are they? Sinkholes.

SAME RESULT, DIFFERENT CAUSES

In April 2014, a Florida resident, Jeremy Bush, lost his brother to a sinkhole. He was awakened in the night by a loud bang. He said, "I ran into Jeff's room and just saw a massive hole." Jeff's room, and other parts of the house, had just dropped into a sinkhole. Jeff's body was never recovered.

In May 2010, a giant sinkhole opened up in a suburb of Guatemala City. The deep hole, which formed an almost perfect cylinder 20 metres wide and 30 metres deep, swallowed a three-storey building.

In 2015, a small sinkhole formed in the backyard of a family home in Illawong, a suburb of Sydney. At 4 o'clock in the afternoon of 15 June, a father and his baby were sitting on the couch when the ground outside suddenly fell away. The collapse was accompanied by a sudden explosion of dirt, which landed all over the back of the house. A 5-metre tree and a washing line were knocked down. Luckily, the family and their neighbours were OK – but they had to be evacuated from their homes.

Three different sinkholes – with three different causes.

SINKHOLE 101

Speaking generally, a sinkhole can range anywhere from a slight depression in the ground, right up to an enormous cylindrical hole reaching down over half a kilometre.

The 2015 Illawong sinkhole was about 15 metres by 15 metres across, with a 3-metre depression. At the other extreme is the world's deepest sinkhole in Chongqing in China. It reaches down 662 metres.

Sinkholes have no natural surface drainage. This means that any water that gets into a sinkhole can't get out via the surface. So the water usually drains downward, into the sub-surface layers.

The common cause of all sinkholes is fairly straightforward. It's just the stuff immediately below the surface shifting to somewhere else.

Depending on how much stuff shifts, you can get anything from a small depression in the surface right up to a circular hole vanishing out of sight into the depths.

So, water does the shifting of the sub-surface material. There are three main patterns.

SINKHOLE CAUSE 1: DISSOLVE LONG-TERM

First, some underground rocks will actually dissolve in water. These include salt beds, gypsum, limestone and other carbonate rocks. This is the situation for the land area in 10 per cent of the whole world, 20 per cent of the USA, or practically all of Florida. (Geologists call this "karst terrain".)

Gypsum is so soluble that if you place a block of gypsum the size of a four-wheel-drive vehicle in a flowing river, it will dissolve away completely within 18 months.

Over hundreds or thousands of years, natural underground currents will slowly dissolve any limestone or gypsum – creating a void. Making the problem worse is that rain is sometimes slightly acidic – so the rock dissolves more quickly. The top of the void gradually migrates towards the surface. Slowly, the surface layer gets thinner and thinner. Solid rock actually turns into a thin bridge, which at some critical stage becomes too weak to support what is above it.

This was the situation in Florida when part of the Bush family's house fell into a sinkhole.

Until 2007, it was compulsory in Florida to have Sinkhole Insurance. Florida is largely underlain by limestone. If you look at a satellite map of Florida, or just fly over it, you'll see it's peppered with many little circular lakes and sinkholes. Practically all the water dissolves outward, from the centre, evenly in all directions.

SINKHOLE CAUSE 2: SHIFT LONG-TERM

The second cause of sinkholes is when the rocks below the surface are made of small grains. They are so small that they can be carried away by persistent, long-term underground water currents.

This was the case in Guatemala City, where a three-storey building just vanished into the sinkhole. The underlying rock was predominantly weak, crumbly volcanic rock, a mix of fine ash, and pyroclastic debris erupted from a volcano in the past. The long-term water currents slowly carried the fine grains away.

SINKHOLE CAUSE 3: SHIFT SHORT-TERM

The third cause of sinkholes is not long-term underground water currents. Instead, it's changes in the short-term water movements above ground.

These changes can be caused by intense rainstorms, or floods, or, rather surprisingly, drought. But most commonly, it happens because we humans changed the water-drainage system that used to be there in the past.

This "change" can range from a burst water pipe to long-term leakage from a broken sewer or storm-water pipe. Worldwide, there is a general trend not to maintain the old infrastructure that was installed decades or even a century ago – and this makes broken pipes more likely. Another major change to the previous drainage system includes covering vast tracts of land with non-porous concrete. Even pumping too much water from underground aquifers has caused sinkholes to form.

In each of these cases, the water can't go where it used to, so it has to find a new path.

The suburban backyard sinkhole in Sydney in 2015 happened after weeks of intense rain, with lots of widespread flooding. After the sub-surface stuff was washed away over several weeks, the last stage happened quite abruptly. One afternoon, without any warning, the surface just slumped down to form a 3-metre-deep basin.

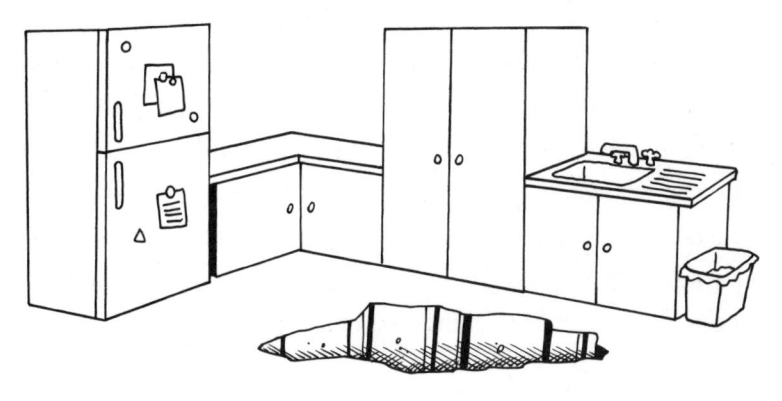

TREATMENT?

But in each of the three main examples, the sinkhole didn't form in less than a minute.

It built up slowly over many years, or weeks. Only the very last stage – the formation of the hole or basin at the actual surface – was sudden.

You might notice a sinkhole close to home by the warning signs of sagging trees or fence posts, doors or windows that don't close properly anymore, or unlikely collections of rainwater.

But mostly, they're hard to find, unless you use ground radar.

Local history helps.

For example, in Australia, the local geology makes Mount Gambier in South Australia very prone to sinkholes. Luckily, the fairly constant water table and dry conditions stop massive numbers of sinkholes from forming. But if the population were to increase from approximately 30,000 to 2.5 million, then sinkholes might well start forming. Other susceptible parts of Australia include: parts of the Nullarbor Plain; the southwestern coastline of Western Australia; Canberra; the southern coastline of the Eyre Peninsula and around the Glenelg River in South Australia; and in New South Wales, Newcastle and, of course, the Jenolan Caves.

And what do you do once you've found one?

Well, filling them in sounds easy. But that sinkhole is a direct conduit to your local water table (or groundwater). If you dump something in a sinkhole now, you could be drinking it in the future. A proper engineering solution could involve filling the sinkhole with rock, cement or grout.

So be on the lookout! If you start to get that sinking feeling, and feel the earth move under your feet, you could be seconds away from being swallowed whole – rather than falling in love.

SINKHOLES

AREAS SUSCEPTIBLE TO SINKHOLES IN AUSTRALIA

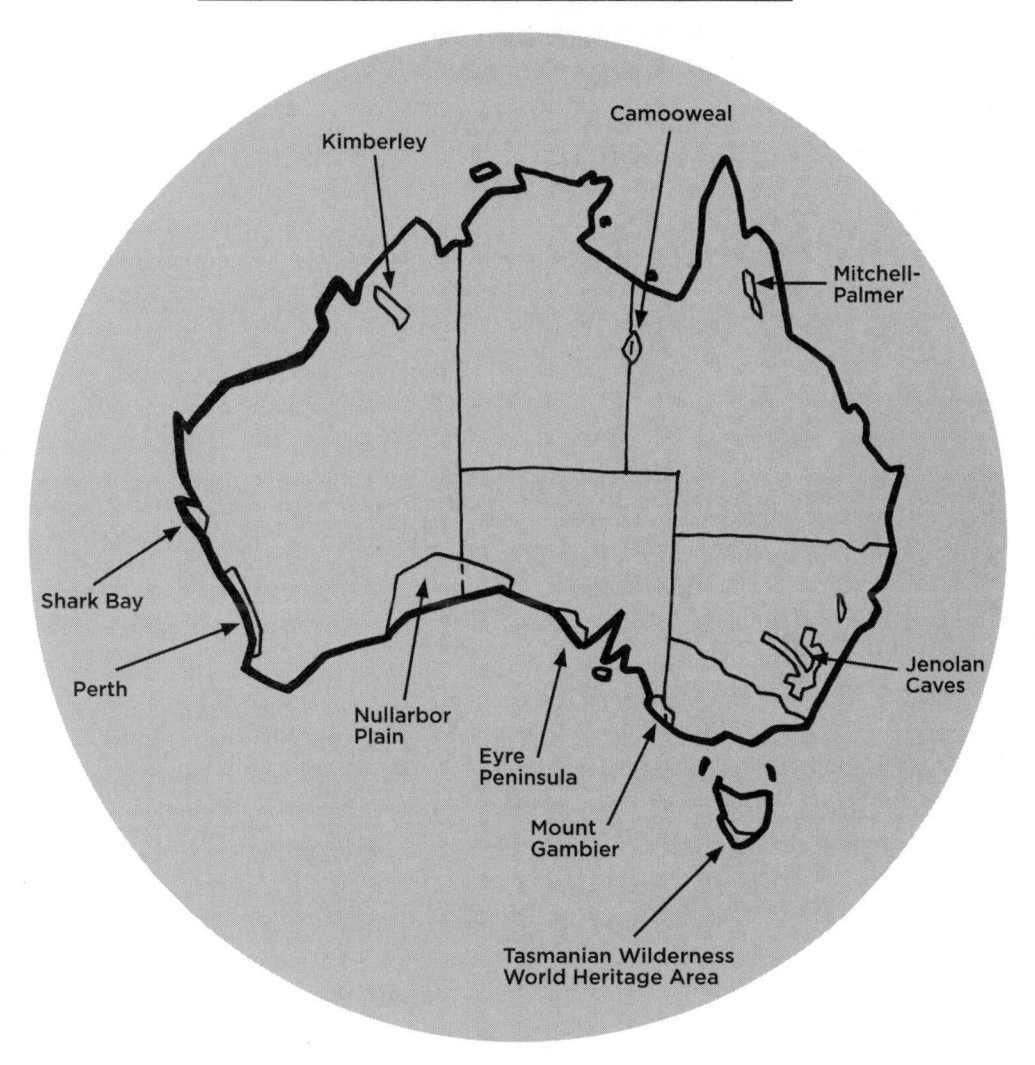

- Kimberley
- Camooweal
- Mitchell-Palmer
- Shark Bay
- Perth
- Nullarbor Plain
- Eyre Peninsula
- Mount Gambier
- Jenolan Caves
- Tasmanian Wilderness World Heritage Area

Collapse or Subsidence

Another way to classify sinkholes is by how quickly they happen.

Cover-collapse sinkholes develop suddenly over a few hours or less. Due to their speed, they can cause catastrophic damage.

Cover-subsidence sinkholes develop much more slowly. The ground very gradually subsides (or deflates), and the change is much less noticeable. These can go undetected for long periods of time.

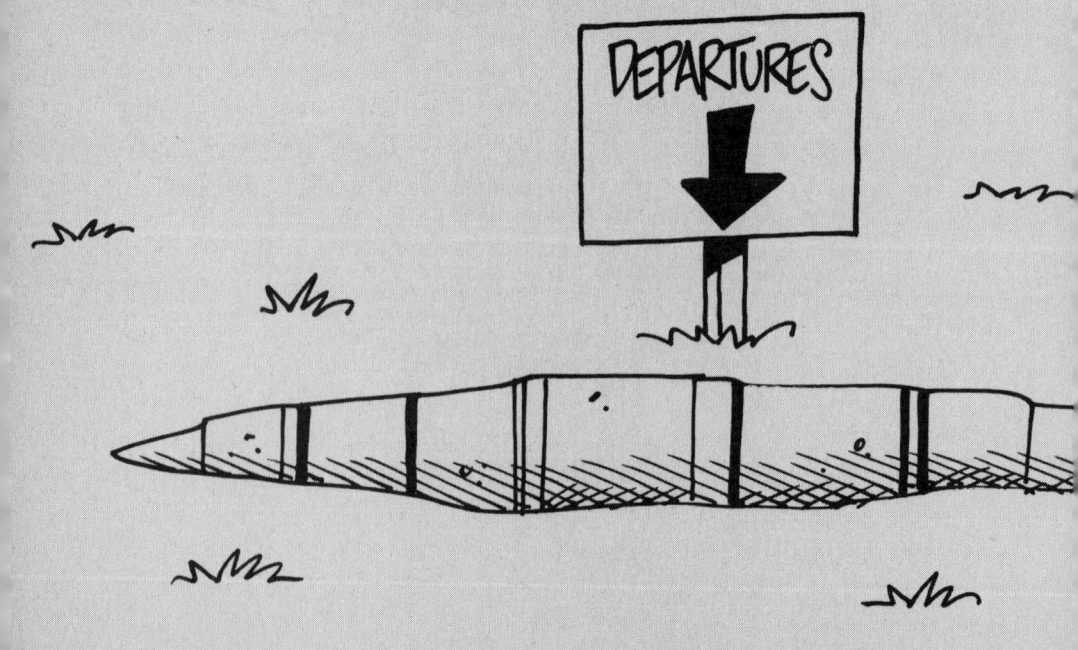

Deep Sinkholes

The deepest known sinkhole is the Xiaozhai Tiankeng (literally, "heavenly pit") in Chongqing in China, at 662 metres deep.

Other deep sinkholes include:

Dashiwei Tiankeng in Guangxi, China – 613 metres;

Red Lake in Croatia – 530 metres;

Minyé in Papua New Guinea – 510 metres;

Sótano del Barro in Mexico – 410 metres.

24

EARTH STOPS SPINNING

THE ROTATION OF the Earth is ever so gradually slowing down. I'm cool with that. But once a school student asked me, "What would happen if the Earth suddenly stopped spinning?" My immediate answer was, "Probably nothing good!" I had to hit the books to get more info. The answer blew me away.

On one hand, once the Earth stopped spinning, you could walk all the way around the Earth at the Equator – and do it entirely on dry land! On the other hand, you would be mostly walking through burning heat or freezing cold. The conditions for Life would change so that most of the planet would – very rapidly – become inhospitable. But there could be a milder "Twilight Zone" between the Earth's new hot and cold faces.

STOP SPINNING?

First, let's get really clear about definitions. Bearing in mind that the Earth will keep orbiting the Sun, what exactly do we mean by the phrase "Earth Stops Spinning"?

There are two theoretical scenarios. In each case, the Earth still orbits around the Sun in a year.

In one case, the Earth's spin slows down to just one spin per year, and this gives it a fixed "Twilight Zone" (more below).

In the other, the Earth stops spinning completely, and forever faces the same distant stars. This gives the Earth a Twilight Zone that takes a year to slowly creep around the planet.

EARTH SLOWS SPIN BY 365 TIMES

In the first scenario examined by Witold Fraczek, the Earth slows its spin by 365 times – so it's still spinning. This places one half of the planet in perpetual sunlight (and baking hot), leaving the other half in permanent darkness (and freezing cold). In between

EARTH STOPS SPINNING

these two extremes is a narrow fixed "Twilight Zone" – where Life could possibly eke out an existence.

In this scenario, our planet would take the same amount of time to spin on its own axis as it would to orbit the Sun – a year in each case. (In the same way, the Moon takes about 27 days to orbit around the Earth, and also takes 27 days to spin on its own axis. This is why we see only one face of the Moon. I wrote about this in my 26th Book, *Please Explain*, in the story "Moon Rotates".)

Weird stuff might happen! The oceans on the hot side might evaporate, and deposit the water as ice on the cold side.

EARTH DOES ONE SLOW SPIN A YEAR, WHICH IS HOW LONG IT TAKES TO ORBIT THE SUN

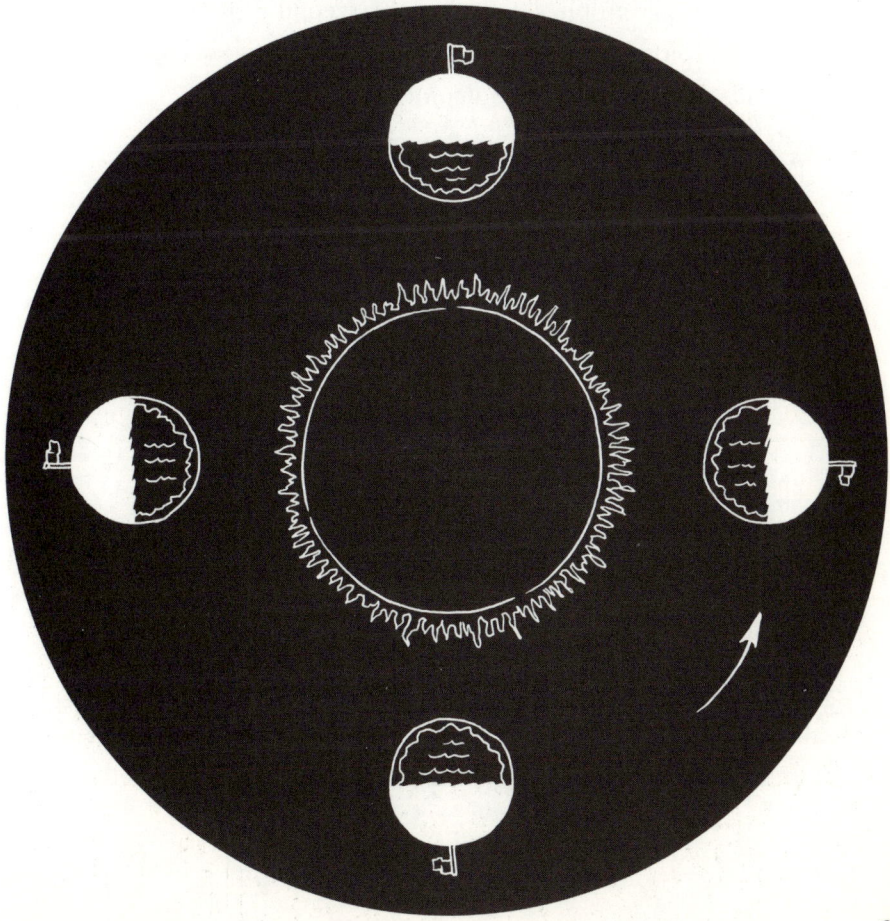

EARTH STOPS SPINNING COMPLETELY

The second scenario that Witold Fraczek explores thoroughly is that the Earth totally slows its spin all the way down to zero.

Earth would still orbit the Sun once per year. If one face of the Earth happened to be facing the Andromeda Galaxy when it stopped, it would continue to do so until our Sun turned into a Red Giant, then a White Dwarf.

The Twilight Zone would slowly creep around the planet over the period of a year, as the Earth tracked its annual orbit around the Sun. At the Equator, the Twilight Zone would travel at about 110 kilometres per day. It would take a very special kind of life to survive these harsh conditions.

To keep things simple, let's choose this second scenario – that the Earth stops spinning completely. (Mind you, the two scenarios are very similar in their ultimate outcome.)

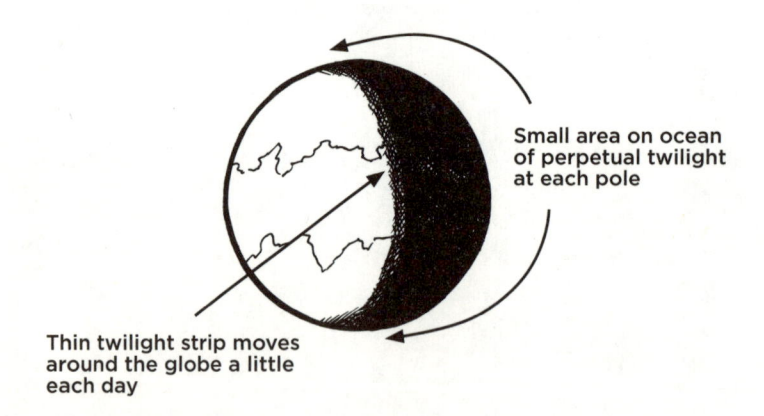

Small area on ocean of perpetual twilight at each pole

Thin twilight strip moves around the globe a little each day

SPIN OF EARTH CAUSED SOLID BULGE

So let's look only at centrifugal force (which should really be called centripetal force – ask a Physicist to explain, if you have a spare hour or so). Centrifugal force is a "fictional" or "pseudo" force that behaves like it is directed outwards, away from the axis of spin (or rotation).

Over several billion years, this force (pushing outwards) has made the Earth a bit fatter around the middle. So today, the diameter of the Earth measured through the Equator is about 42.6 kilometres more than measured through the Poles. (I wrote about this, and how Mount Chimborazo in Ecuador is further from the centre of the Earth than Mount Everest, in my 23rd book, *Great Mythconceptions*.)

But this bulge in the solid Earth took billions of years to slowly develop. This is because the solid matter moved only very slowly in response to the outward force caused by the spin of the planet.

SPIN OF EARTH CAUSED LIQUID BULGE

Compared to the solid crust, the liquid water in the oceans is far more mobile and responsive to forces.

The Earth's spin has actually pushed the oceans outwards at the Equator. The water has gone "uphill" a long way – about 8 kilometres. In other words, at the Equator, thanks to the spinning Earth, the water has been pushed up some 8 kilometres higher than if the Earth had no spin. However, on the entire 40,000-kilometre circumference of the Equator, the deepest part of the oceans is only about 5.25 kilometres. (This deep spot is southwest of Kiribati Island in the Western Pacific Ocean.)

So if you take away the spin (which provides the force pushing the water uphill), then all the water in the oceans starts flowing away from the Equator, and towards the Poles.

WATER FLOWS TO POLES

As the waters retreat, small regions of terra firma on and around the Equator would rise up.

Eventually, there would be a huge megacontinent wrapped continuously around the Earth at the Equator. You could travel around the Earth on the Equator and stay entirely on dry land (ignoring the freezing cold on the night side, and the searing heat on the day side). The lowest point on your journey would be about 2.75 kilometres above "sea level", southwest of where Kiribati Island used to be.

There would be two totally disconnected polar oceans on each side of the Equatorial megacontinent. This is very different from today, where the Pacific, Atlantic and Indian Oceans are all connected to each other by the Southern Ocean.

In the north, Canada would be entirely underwater. And roughly following the line that defines the border of current-day USA and Canada, all of Greenland, as well as the northern plains of Siberia, Asia and Europe would be underwater. But parts of Spain (closer to the Equator) would stay above water.

On the southern side of the Equator, the new expanded Southern Ocean would start roughly on a line running through current-day Canberra. Africa would be joined to the island of Madagascar, while Australia, Papua New Guinea and Indonesia would be one continent.

WATER FLOWING TO POLES

BEWARE YE KEVIN COSTNERS
HERE BE MONSTERS

Polar Oceans at Different Levels

It turns out that the underwater basin around the South Pole is much bigger than the one around the North Pole. (After all, the North Pole is closely surrounded by land, while the South Pole is a long way from the land masses of Africa, South America and Australia.)

So the new Southern Ocean would be lower. It's a bigger "bowl" with a greater capacity. Its sea level would be about 1.4 kilometres lower than the sea level of the new Northern Ocean.

EARTH WILL SPIN

Luckily for us, the Earth will not stop spinning. The Earth is so massive that it would take energies of the order of another planet sideswiping us to quickly change the length of day.

Smaller forces, over eons of time, can change the length of the day – but very slowly, and only by tiny amounts.

Our spinning Earth is slowing down (mostly due to the friction of the ocean tides). Over the period of a century, the day drifts to be about 2.4 milliseconds longer. So billions of years in the past, the Earth spun faster. A day is the time it takes the Earth to do a complete rotation on its own axis, and a year is each journey around the Sun. Around 400 million years ago, there were about 40 extra days in each year. Back then, the day was shorter, because the Earth spun faster.

Billions of years in the future, the Earth will spin more slowly. The outward force at the equator will be less, and so our planet will have a smaller bulge.

The Earth's shape will be closer to a sphere, and it will lose its bigger waistline. The Earth will be able to pull off what the Dieting Industry fails at 95 per cent of the time.

"Solid" Earth?

The Earth is not actually solid.

There is the so-called "solid" crust on the surface – about 6 to 100 kilometres thick. (It's a huge variation, with the crust is thickest at big mountain ranges.) But underneath that is red-hot molten rock (the mantle) reaching down to about 2900 kilometres below the surface. Under that is a layer of liquid iron. In turn, this liquid iron surrounds a ball of solid iron at the centre.

Planet "Earth" or "Water"?

By itself, the Pacific Ocean is larger than all the land areas on Earth combined. About 70 per cent of the surface of our planet is covered with water, not land.

So our planet should be called "Water", not "Earth".

Spin versus Gravity

The spin of the Earth has given us today's 8-kilometre-high bulge of water at the Equator.

It could have been a bit higher than that, but Gravity made it a little smaller. The Poles are about 21 kilometres closer to the centre of the Earth than the Equator. So the gravity is ever so slightly stronger at the Poles. This slightly draws the water away from the Equator.

25

CANE TOADS CONFIRM CONCEPTION

TODAY, EARLY PREGNANCY testing is very simple. Head to the chemist, buy a Home Pregnancy Test Kit, wee into a jar, dip the test stick into urine, and watch and wait.

It took several thousand years to make something as complicated as testing pregnancy so simple. Before Home Pregnancy Test Kits were freely available, we had tried burning urine-soaked cloth, injecting urine into mice and rabbits – and, for the uniquely Australian touch, cane toads.

PREGNANCY TESTING HISTORY

Back in the time of the pyramids, the ancient Egyptians poured the urine of a potentially pregnant woman onto grains of wheat and barley. If they sprouted, she was pregnant. The type of grain that sprouted was meant to show the sex of the baby. This test was totally unreliable.

Dipping a ribbon into the woman's urine and then burning it, as they did in the Netherlands in the 17th century, didn't work either.

Even in the early 20th century, there was no reliable test for pregnancy.

Symptoms of pregnancy were recognised, of course – such as missed periods, nausea and vomiting, breast tenderness and food craving.

Skilled midwives and obstetricians looked for clinical signs of pregnancy. They checked for enlargement of the uterus, and sometimes could see colour changes in the cervix (the bottom of the uterus) due to an increased blood supply in early pregnancy. Other clinical signs included extra pigmentation of the skin, such as darkening of the nipples.

PREGNANCY TESTING – MICE AND RABBITS

Around 1928, the German gynaecologists Selmar Aschheim and Bernhard Zondek discovered a mystery hormone in the urine of pregnant women. This hormone (beta-human chorionic gonadotropin, or βhCG) was not isolated until the 1950s. But they thought that they could use this still-unidentified chemical to test for pregnancy. They injected the urine of potentially pregnant women into mice, twice a day for three days. They killed the mice 100 hours after the first injection and examined their ovaries. If they could see new blood vessels on the ovaries, they were 99 per cent sure the woman was pregnant.

This test soon evolved into the Friedman Test, developed in 1931 by American researchers Maurice Friedman and Maxwell Lapham. Early-morning urine was injected into the ear vein of a virgin female rabbit, aged 12 weeks or more. The rabbit was killed after 36 hours. Characteristic changes in the rabbit ovaries showed that the woman who supplied the urine was pregnant.

PREGNANCY TESTING – CANE TOADS

One problem with using mice and rabbits was that they had to be killed to make a diagnosis. But toads are different from mammals, and didn't have to be killed.

Cane toads had been introduced into Australia in 1935. By the 1950s, they were a real pest in Cairns. In classic Aussie style, Louis Tuttle and Bill Horsfall from the Cairns Hospital built on the work of others and developed a pregnancy test using "surplus" cane toads.

They isolated male toads from female toads for a few weeks. This made sure the male toads were not generating any sperm. At 9:30 a.m., a potentially pregnant woman's urine sample was injected into the back of the male cane toad. (The woman's early-morning urine sample would contain the highest levels of βhCG.) The toad was then examined at 3 p.m. and 5 p.m. on the same day. If the male toad produced sperm, the pregnancy test was positive. The male cane toad did not have to be killed, and could be used again and again. Soon cane toads were being flown out of Cairns airport to hospital laboratories across Australia. The only downside was that this didn't really reduce cane toad numbers at all.

PREGNANCY TEST – IMMUNOLOGY

In 1960, an immunological test for pregnancy was developed. This is the type of test we use today. It became available over the counter in Canada in 1971, and quickly spread around the world. However, its introduction into the US was delayed until 1977, because of "concerns" that it might encourage immorality in women, and that women would be unable to interpret the results correctly without the intervention of a medical doctor!

In the mid-20th century, with cane toads, mice and rabbits needed to detect pregnancy, and storks to deliver babies, you were virtually running a bizarre baby-making mini-zoo.

26

CHRONIC LATENESS

SOME OF US are often late for our appointments – and to be brutally and painfully accurate, I'm one of those nearly-always-late people. The Land of Chronic Lateness is not a happy place to be. But even when Lateness is chronic, it is still possible to break this lingering habit.

After all, time is a scarce and a valuable resource. You can always make more money, but you can't make more time. It makes obvious sense to use your limited time efficiently.

LATENESS OK OR NOT OK?

One way to look at Chronic Lateness is the effect that it has on others. There's so-called "OK Lateness" and "Not OK Lateness".

Suppose that you're going to the movies alone and you are running late. Maybe you miss the beginning of the latest movie, and you probably get a lousy seat – but you are the only one affected. That's OK Lateness.

But Not OK Lateness is when you affect others – when the meeting or the dinner can't start without you. If you've kept 15 people waiting for 8 minutes, you've just wasted two hours of other people's time.

Apologising over and over is not going to make up for wasting two hours of other people's time. And waste more of their time by apologising!

It turns out that Not OK Late people can be further divided into another two subgroups.

The first group doesn't care about being late, and doesn't feel bad. Unfortunately, sometimes you simply have to avoid these people, or at least, keep your dealings with them to a minimum.

The second group does care and does feel bad about keeping you waiting. This is the group that can be helped.

Passage of One Minute

An interesting aspect of Chronic Lateness is that we all experience the passage of time in our own individual way.

One study asked people to estimate how long it took for a minute to pass.

It appeared that people who are achievement-oriented, fast-paced, highly strung and occasionally hostile (classic Type A) perceived a minute as passing in 58 seconds. Not surprisingly, Type A personalities tend to be punctual.

But the more laid-back and relaxed people (classic Type B) thought that it took a comfortable 77 seconds for that minute to float past them.

PERSONALITIES PRONE TO LATENESS

Dr Linda Sapadin is a Fellow of the American Psychological Association, and she specialises in Time Management. She has identified four types of personalities that are especially prone to being chronically late. They are the Perfectionist, the Crisis Maker, the Defier and the Dreamer. (Let me acknowledge here that while Dr Sapadin's approach has much to commend it, it is a slightly simplistic approach. It makes no specific allowances for people who have troubles with executive function, i.e. planning ahead, regulation of own emotions and impulses, and so on. However, within those limitations, Dr Sapadin's concepts are a good First Approximation to helping fix Chronic Lateness.)

PERSONALITY 1: PERFECTIONIST

The Perfectionist simply can't leave home until the dishwasher is packed (perfectly, of course) and set running. Furthermore, everything else has to be perfect, including their appearance, and the project they're presenting.

Unfortunately, their concept of perfectionism doesn't extend far enough for them to realise that being late for the meeting rules out the possibility of a perfect presentation.

So they need to learn to see the Big Picture, and to realise that small details can be left till later. Maybe they can try to leave the dishwasher unstacked, and arrive early enough to set up the presentation before people come into the meeting room.

PERSONALITY 2: CRISIS MAKER

Another personality type is the Crisis Maker.

They might not specifically want to be always late, but the pressure and the adrenaline rush gives them a nice thrill that they keep chasing. They can't start on something until just before the deadline, because they think they can't function well unless they are fully hyped up. (I have a sneaking suspicion this might be me. Perhaps my motto, "If it wasn't for the last minute, nothing would ever get done" is "unhelpful"?) These people actually prefer to be desperately rushing to get to their next appointment than to stroll calmly into the building 5 to 10 minutes early.

They might be better off getting their adrenaline from physical activities, rather than from the terror of an approaching deadline.

The Take-Home Message for these people is – be a Thrill Seeker/Crisis Maker on your own time, not on other people's time.

PERSONALITY 3: DEFIER

The Defier feels that they have to stand up against the broad authority of our existing societal constructs that tell us what to do, and when to do it. An "easy" way to Fight The Man is by being late.

This person might be able to realise that they are always reacting to what they see as the oppressive forces of society. They don't make the first move – they are always on the back foot.

So instead, they could "act" rather than "react". They could face society on their terms, not society's terms.

And part of that would be to unnerve the Running Dog Lackeys of the Oppressors of the Honest Hard-Working Proletariat by being fully prepared, and even on time. (Just a thought . . .)

PERSONALITY 4: DREAMER

Finally, we have the Dreamer.

Dreamers live in a different reality. They are bizarrely confident that they can have a shower, pack up all their luggage in the hotel room, take the elevator downstairs, wait in the queue at Reception, check out of the hotel and be travelling in a taxi to the airport in a total of 10 minutes.

They think that time works differently for them. So they "see" travel times as really short, and imagine that it's perfectly reasonable to fit five jobs into a five-minute window. Often these people can be brought back to a real time frame via various travel apps on a smartphone. And if they can see the reality of the situation there on the screen (31 minutes to the airport, not 10), then they have to reset their schedules.

SOLUTIONS

On average, the "chronically late" will underestimate how long tasks or events take by around 40 per cent. That is surprisingly large.

It might sound odd to say it, but some people need to re-learn how to judge the time. They may well have learnt in kindergarten but if they did, they kind of forgot when they grew up. It can be very difficult to change the habits of a lifetime but you may as well start somewhere. An excellent set of solutions comes from an article, "Can You Cure Chronic Lateness?" in *The Atlantic* by Li Zhou.

On each occasion you are late, take the time to work out why this happened. One tactic is to find the "pain points" when you are late, and pick on just one of them. For example, your bus might take 30 minutes to get you to your destination – but your waiting time for the bus could be 10 minutes. So allow 50 minutes (not the hopefully optimistic 40 minutes or the totally irrational 30) for the whole journey.

Start small, succeed with that one pain point, and repeat the winning strategy over and over to lock it in. Then add a new pain point.

Repetition and consistency are essential. Along the way, you'll start getting positive feedback, such as benefitting from having the time to gather your thoughts and analyse situations. You might even arrive at the party before all the food has been eaten!

Another good trick is to wear an Old School wristwatch. All they do is tell the time. Even though a smartphone is usually more accurate than a wristwatch, it can easily distract you with all those texts, notifications, Snapchats, Instagram updates, tweets, etc. continually flooding in. An added advantage is that it's quicker to read a wristwatch than to pull a smartphone out of your pocket or bag.

Definitely avoid split-second timing. Many people who are chronically late hate to "waste" time waiting for others. So, OK, get around that – bring a book (either paper or an ebook).

If you write down daily plans, you get an overview of your time and can see what is reasonable, and what is not. It's also good to have blocks of time each day with nothing in them. This gives you a chance to be creative.

You've spent decades locking in the habits of Chronic Lateness. The bad mental pathways have been repeatedly reinforced. It will take months (not days) to remove them – and then build up and strengthen brand-new mental pathways.

But it's never too late . . .

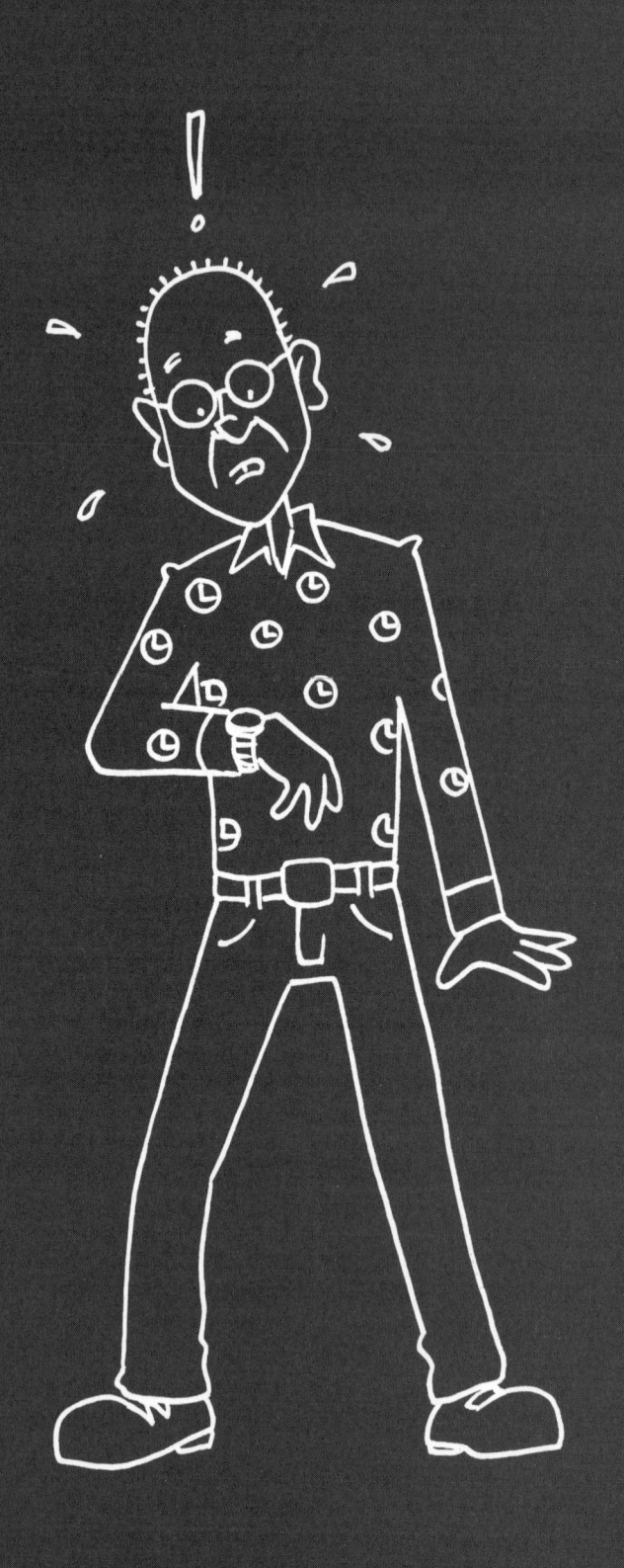

Are You Always Waiting for Others?

(I learnt a lot from the *Elle* article, "How to Stop Being Late for Everything".) Learn from Professionals. Follow the example of dentists and doctors. Ring the other party, and ask if they are on time for your appointment.

Time for Tough Love. Let the other person know that you feel disrespected, and that you have been inconvenienced by their lateness. Tell them if they keep you waiting for dinner for 15 minutes, it's their shout. If it's a business appointment, tell them you'll wait 15 minutes – otherwise, you'll have to leave as their tardiness will make you late for your next appointment.

What about White Lies? Sorry, but sometimes they are necessary. Tell the latecomer that the movie starts at 7:30 p.m., instead of 8:00 p.m.

Be Strong. Don't take it personally. They are late for everything and everyone – not just you. It's not personal – it's (their) business as usual.

27

PB/5
PEDESTRIAN BUTTON

↑

IN 1884, THE flamboyant author Oscar Wilde wrote, "I have found that all ugly things are made by those who strive to make something beautiful, and that all beautiful things are made by those who strive to make something useful."

It's a shame he never saw the PB/5.

Never heard of it? Luckily for you, I'll fix that problem right now.

The PB/5 is the standard pushbutton switch that Australian pedestrians press when wanting to cross the road at a traffic light – and I love it to pieces. It's so useful that it has become totally inconspicuous. Most pedestrians never give the PB/5 a second thought – they just automatically push it.

But, you might ask, what is the Big Deal? What is there to know about a simple push button switch?

SWITCH 101

One of my techno friends had plenty of answers to that question. He gave lectures to Audio Engineering students about Switches and Buttons. After I attended one of these lectures, I saw switches and buttons in a whole new light.

SWITCH OFF: No electricity flow

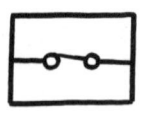

SWITCH ON: Electricity flows through a "contact"

An "ideal" or "perfect" switch should have contacts (usually metal) that "switch" (or connect) electricity instantly from one contact to another contact. In addition, any moving contacts should not physically "bounce" on and off when they change location, they should have zero resistance, they should have no effects on the voltage or the current, etc. But real switches fail in each of these attributes (to some degree). Every switch is a compromise.

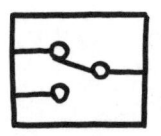

**This switch can send electricity to
either of two destinations**

Switches can be turned on by a human (like you) or by some other stimulus. This stimulus could be vibration, the angle of tilt, the level of a fluid, the air pressure, the turning of a key, or even a magnetic field (the reed switch).

There are many different types of switches all around you. They include circuit breakers, microswitches, rocker switches, in-line switches, surface mount switches, wafer switches, toggle switches, reed switches and the one that operates your house lights – the wall switch.

There really is a lot to know about switches. If you look around your house, you'll quickly see that most switches take your device from "on" to "off", and back again. While most switches will "toggle" into (and stay in) whatever mode you place them, some are "momentary" – they operate only while you still hold the button.

But switches get way more complicated once you get into computers and their associated hardware.

Not-So-Humble Wall Switch

Back in the old days, we had tungsten incandescent light bulbs. They were terribly inefficient. They should really have been called Heat Bulbs, because they made 10 times more heat than light.

But they had another problem. Their high steady state current (when they were up and running) was bad enough (about half an ampere). But they had an "inrush" or "warm-up" current that was 10 times larger than the steady state current. (I wrote about the myth that turning on a light bulb used enough electricity to run the bulb for 30 minutes in my 23rd book, *Great Mythconceptions*, in the story "Switch on Light Bulb". Yes, the inrush current was 10 times greater, but not tens of thousands of times greater, as the myth claimed.)

This inrush current happened immediately after the wall switch was turned on. At this stage, before the bulb lit up, the tungsten filament was cold, and so it had low resistance. But as the filament warmed up and started glowing, the resistance increased and the current quickly dropped to its steady state.

The wall switches had to be specially designed to handle this temporary massive current – not just the much lower steady state current.

So even the humble wall switch for the light bulb had its own special subtleties.

Irrational Switches

I have a Keyboard, Video, Mouse (KVM) Switch that is really not intuitive to use. The typical KVM lets you use a single Keyboard, Video Monitor and Mouse with two (or more) different computers – you just press the switch to select which computer you want to use. For a few months, I had an incredible amount of trouble with this KVM Switch. I could not make it work reliably, and I didn't know why.

   Eventually I worked out that it performs three different operations, depending on how long you hold it down for: 0 to 3 seconds, 3 to 15 seconds, or longer than 15 seconds.

   How did I find this essential timing information? By reading every word on every page of the manual. Frustratingly, it was hidden on page 45 of a 180-page manual!

Impotent Button Pushing

In big cities, most pedestrian switches don't care how often you press them – whether it's once or a thousand times. (I wrote about this in my 26th book, *Please Explain*, in the story "Button Pushing".)

They still act upon your signal – but only sometimes. The rest of the time, they are programmed to ignore you.

In the Central Business District of some Australian cities, the traffic light push buttons have been specifically set to not work at certain busy times. These busy times are typically 7 a.m. to 7 p.m. Monday to Wednesday, and an extra two hours on Thursday to Saturday (7 a.m. to 9 p.m.). The rationale is that during these busy hours, the road traffic is relatively constant, and the pedestrian crossings would normally be in continuous use.

So the lights change on their own internal program, which is totally unrelated to your button pushing.

However, the traffic light push buttons do work to trigger the light change outside these hours, including all day on Sundays.

PB/5 101

Back in the 1960s, public provision of services for people who are deaf and blind in our society was slowly increasing. A few audio-signalling pedestrian buttons had been installed, but they had confusing non-intuitive features, and weren't very sturdy.

In 1976, in NSW, the then-Department of Main Roads commissioned a team to come up with something better. The team included industrial designer David Wood (Nielsen Design Associates), acoustic consultant Louis Challis (Louis A. Challis and Associates) and engineer Frank Hulscher (Department of Main Roads).

Now if you're designing a switch for pedestrians to use when crossing the road, you need something robust, elegant and easy to use.

The Australian PB/5 has all this under control. Its official name on the patent is "audio-tactile pedestrian push button signalling system". It started appearing in Australian streets in 1984, and has since been exported to the USA, Ireland, New Zealand and Singapore.

You've seen (and probably not noticed) the PB/5 a thousand times.

It's a 25-centimetre-tall epoxy powder-coated diecast aluminium box with safe rounded edges, and it's bolted onto a traffic pole at a pedestrian crossing. (Some later models used high-grade engineering plastics). The PB/5 has been designed to look similar to a traffic light. (Now isn't that a lovely subtle design touch?) The circular upper half has a distinctive large white arrow on a blue background. It has a smaller raised arrow inside it, so that people who have low visibility can feel it. The bottom half has a generously sized concave circular stainless steel button – easy to hit.

It's also weather- and vandal-proof.

**THE PB/5
PEDESTRIAN BUTTON**

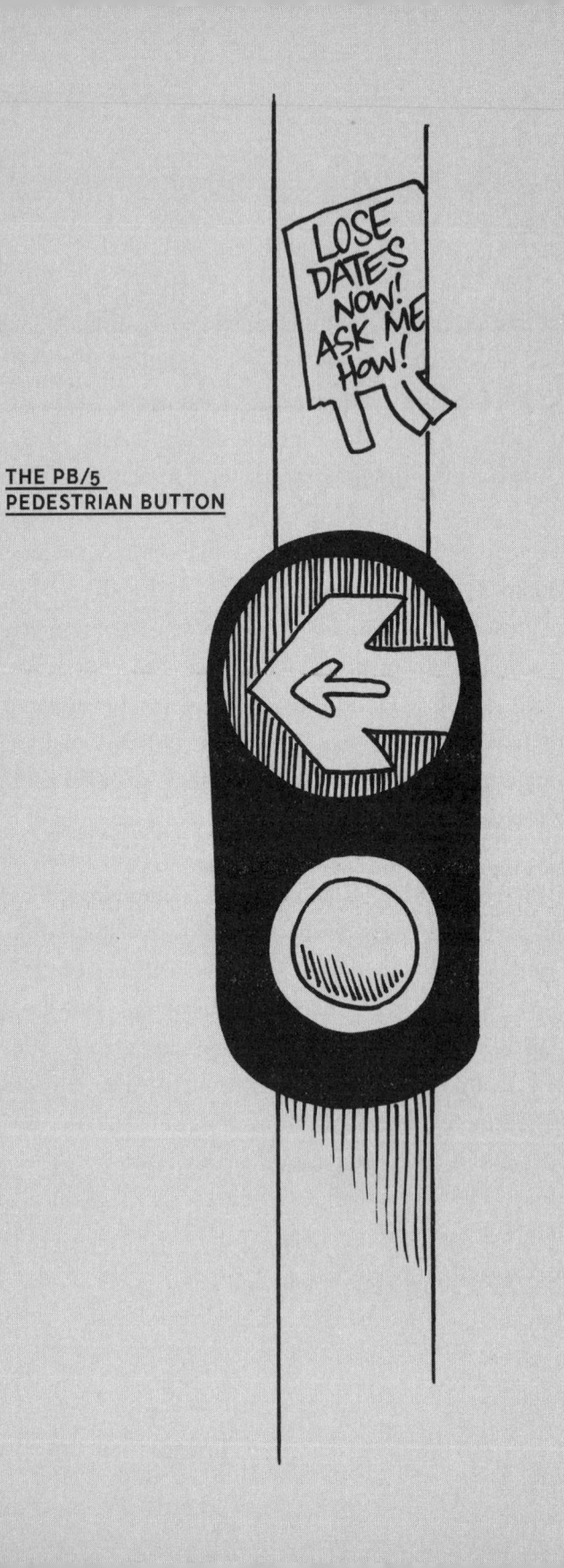

SOUND OF MUSIC

The sound mechanism inside PB/5 is officially called the Two Rhythm Buzzer. It generates only two sounds – the slow "chirp" and the urgent "tick-tock-tick-tock". But it also "creates" a single third sound, the "kapow!", by a clever transition from the slow chirp to the urgent tick-tock.

These sounds are all part of the elegant design.

First, the slow chirp (the "don't walk" signal) alerts you once every two seconds that this is the time to wait. It also "calls" you to the location where you should wait. The patent describes the "chirp" as the "Audible Locator Signal" – so that vision-impaired people can easily find the PB/5. The frequency has been specifically chosen to be easily identifiable in city traffic noise. (Each chirp is a 1 kilohertz square wave.)

Second, the single "kapow!" noise (the "change tone") alerts you that something has indeed changed. (It's a square-wave burst starting at 2 kilohertz, which drops to 500 hertz over a 50 millisecond period.) It tells you that it's time to pay attention – something has changed.

Third, the rapid tick-tock-tick-tock sound (the "walk" signal, repeated 8 to 10 times per second) tells you to get moving. (Each tick is a 500 hertz decaying sinewave.)

VIBRATING TOUCH PANEL

In the upper half of the PB/5, the artwork of the white arrow on the blue background is permanently anodised onto a circular metal plate. This metal plate is joined to a "transducer". As part of the elegant design, this single transducer does three jobs.

First, the transducer acts as a loudspeaker to make the three different sounds.

Second, it also does the exact opposite of a loudspeaker. It acts as a microphone that listens to the ambient city noise. This then controls an internal amplifier to make the loudspeaker louder or softer. In the busy city on a noisy day, it will become louder so that you can hear it above the city noise. But in the quieter suburbs, late at night, it will become much softer – still loud enough to be heard, but not to be irritating.

Third, the transducer can vibrate the little white raised arrow (inside the bigger white arrow) in different modes that correspond to the different sounds. This means that people who are hard of hearing can touch the metal plate and know whether to wait, or walk. This is the Vibrating Touch Panel.

The Vibrating Touch Panel is also a Braille Direction Arrow. It can point straight ahead, left, or right – to tell you which way to walk.

TOUGH SWITCH

One design attribute of every switch is the Mean Time Before Failure (MTBF) – how many operations it will carry out before it fails.

The PB/5 has a MTBF of many millions. I've seen people kick the stainless steel button – and it easily survives.

This part of the overall design is very clever.

First, the stainless steel button slides in and out on a robust mechanism.

Second, there is no direct physical link between the sliding stainless steel button and the electrical contacts. Instead a magnet is attached to the button. When this magnet slides past a reed switch, it turns it on – with no direct physical contact. The reed switch then closes electrical contacts in the same way every time – regardless of whether you gently pressed the big button, or kicked it as hard as you could.

PB/5 PEDESTRIAN BUTTON

The PB/5 has many classic design attributes – subtlety, safety, form and function, and innovation. As switches and push buttons go, it's way ahead of the pack.

I'm pretty sure that Oscar Wilde would have loved the usefulness – and beauty – of the PB/5 Audio-Tactile Pedestrian Push Button Signalling System as much as I do.

OSCAR WILDE

28
BEER "BEATS" CANCER?

THERE ARE BENEFITS from eating and cooking meat. Meat is loaded with protein, B vitamins, iron and zinc. Cooking kills most, if not all, bacteria. The cooked meat is a lot easier to digest, and the texture and flavour improve enormously.

On the other hand, red and processed meats have been linked to increased health risks.

But marinating the meat before cooking can reduce some of these risks.

MEAT AND CANCER

There is some convincing evidence for a link between eating red and processed meat, and increased risk of colorectal cancer. For every 100 grams of red and processed meat eaten on average per day, your bowel cancer risk increases by 14 per cent. (Unprocessed white meats are lower risk.)

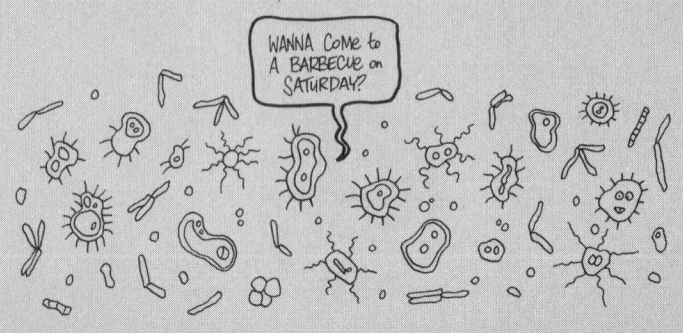

The problem can be the saturated fats in some meats – and also the cooking method. High-temperature methods of cooking (like grilling and frying) are associated with increased risk of polyps of the colon and rectum. Carcinogenic chemicals can also be formed when fat drips onto a heated surface and then burns.

It seems logical to reduce nasty carcinogens in your meat. A surprising way to do this is to marinate meat before you barbecue it.

MARINADE

Three types of marinade were tested – Pilsner beer, wine and de-alcoholised wine. In each case, the marinade had the option of including herbs and spices. So 100 millilitres of marinade typically contained 2.8 grams of ginger, 2.9 grams of garlic, 0.4 grams of rosemary, 0.25 grams of thyme and 0.1 grams of red chilli pepper.

The meat was marinated for four hours, and then pan-fried in a Teflon-coated pan for three minutes on each side.

All of the tested marinades reduced carcinogens in the cooked meat – to some degree. Beer marinades were more efficient at reducing carcinogens than white-wine marinades. In each case, the reduction was improved further by adding herbs.

So if you're eating meat, make it lean, white and unprocessed – sprinkle those herbs and marinate the carcinogens away.

29

MARIJUANA FOR MEMORY & LEARNING?

EXTRA, EXTRA, READ all about it! These headlines went for maximum impact.

"Dope Could Help Make Old Mice Less Dopey"

"Daily Dose of Cannabis Extract Could Reverse Brain's Decline in Old Age, Study Suggests"

"Marijuana May Boost, Rather than Dull, the Elderly Brain"

BACK TO THE SOURCE

Was this real? Could marijuana boost the elderly brain? And what would it do to the young brain?

I went straight to the original peer-reviewed paper in *Nature Medicine*. This is a pretty serious journal – no fake news here.

The paper had the title "A Chronic Low Dose of Delta-Nine-Tetrahydrocannabinol (THC) Restores Cognitive Function in Old Mice".

The scientists were from the Institute of Molecular Psychiatry at the University of Bonn in Germany.

Their experiment included mice that were either young (2 months), mature (12 months) or old (18 months).

Normally, with mice, you would expect that learning and memory performance would decline with age.

THE STUDY SHOWED THAT . . .

In this study, all the mice were implanted with a continuous low-dose liquid-infusion pump. For 28 days, some received a placebo, while others were given continuous low doses of THC – one of the major active psychedelic ingredients in marijuana. (THC is just one of at least 85 cannabinoids in cannabis.) The dose of THC was set at 3 milligrams per kilogram of body weight per day. It was (apparently) not enough to make the mice "high".

MARIJUANA FOR MEMORY & LEARNING?

The mice were tested during this 28-day window during which they were infused, and for another few weeks afterwards. The tests covered various intellectual tasks — especially memory and learning skills. They included Long-Term Spatial Memory (finding a safe platform in a water maze), Novel Object Location Recognition (remembering new objects), and Partner Recognition (recognising a mouse they had met before).

First, let's look at the placebo — no THC for any mice. This was their "normal" state. Yep, the young mice did well, and both the mature and the old mice performed worse than the young mice. No surprise at all.

Second, the effect of THC on younger mice. They performed worse. Again, no surprise.

Third, the effect of THC on mature and old mice. They got smarter. THC let them match the intellectual performance levels of healthy younger mice (who were not on THC). This was astonishing.

These good effects lasted for a few weeks after the THC pumps had been switched off.

Fifty days after implantation of the pumps, the scientists examined the brains of all the mice. In particular, they were looking at the synapses (or nerve connections), as well as gene expression, in the hippocampus — a part of the brain strongly involved with memory.

The brains of the older mice that had taken THC had improved. They looked just like the brains of the healthy younger mice that had not taken THC.

WHY DOES CANNABIS WORK ON HUMANS? PART 1

But why does THC affect mice and human brains? After all, THC comes from a plant, and we're made from meat.

Well, to answer that, consider a pretty red poppy from Turkey. It gives us opium. How come this chemical extracted from this poppy is really good at relieving some kinds of pain in humans? What's the link between plants and humans?

In the 1980s, we discovered that humans make their own natural opiates. We call them "endorphins". Endorphins do their work by acting on our natural Endorphin System. It's just a coincidence.

It turns out that opiates, which are very similar to endorphins, can affect us humans by acting on our natural Endorphin System. It's just a happy coincidence that plant chemicals work on humans to relieve pain.

WHY DOES CANNABIS WORK ON HUMANS? PART 2

About a quarter of a century ago, we discovered that humans also have our own natural EndoCannabinoid System (ECS).

THC is very similar to endocannabinoids that our brain naturally makes. Yes, another pharmacological coincidence (between a plant and us).

These endocannabinoids operate within our natural Endo-Cannabinoid System to help with many functions in our bodies, including learning and memory. These functions include the growing of new nerves, phosphorylation, chemical homeostasis, regulation of type II immune response, inflammatory response, transmission of nerve impulses, programmed cell death and much, much more. (I discuss the growing of new nerves, neurogenesis, in the story "Childhood Amnesia" on page 184. Programmed Cell Death is in the story "Suicide Cells, Apoptosis" in my 19th book, *Fidgeting Fat, Exploding Meat and Gobbling Whirly Birds*.)

Our ECS develops gradually when we are children, becomes very active in adolescence, and then tapers down. As we get older, our natural ECS degrades with time.

So here is a guess.

In older mice, perhaps long-term low-dose THC rejuvenates their natural EndoCannabinoid System. And this might be how it restores some intellectual functions in elderly mice.

YES, BUT . . .

Let's not get too far ahead of ourselves.

Usually, we test drugs on mice hoping to one day use them on people. But there are some issues here. First, mice are not humans. In fact, the brains of mice are very different from the brains of humans. Mice have very high densities of THC receptor sites in the basal ganglia and cerebellum, while humans have much lower concentrations in these sites. As another example of the limitations of testing mice, we have found about a thousand treatments that help with Alzheimer's Disease in mice. How many of them work in humans? Maybe three, and that's a Big Maybe.

Second, a 28-day period of infusing THC is a very short time. One lousy month is not long enough to make long-term conclusions. What happens after three months? Do the older mice unexpectedly go into an abrupt and irreversible decline?

Third, there weren't a lot of mice in the study – tens, rather than hundreds, of little rodent subjects.

Fourth, the results of what THC did to the mice showed a wide scatter. In other words, some older mice did really well, while others did no better than not having THC. But yes, on average, the older mice improved on THC. (Variations in individual response to drugs are common. In cancer patients, cannabis can improve appetite in some, but cause nausea in others.)

So for our geriatric population, there are lots of caveats with this story. It's way too early to start trying to self-medicate memory loss, possibly caused by dangerously low blood levels of THC. (But what kind of "catchy" headline would that be?)

One of the authors of the paper, Andrés Ozaita, wrote, "If we can rejuvenate the brain so that everybody gets five to 10 more years without needing extra care then that is more than we could have imagined." The team is planning to test the effects of THC in elderly adults with mild cognitive impairments. But at what age – bearing in mind that mice and human brains are very different – should trials in humans begin? Furthermore, to parallel the results in mice, the trial would need 3–4 years of sustained THC in humans.

And the younger mice? Their EndoCannabinoid System was already very active. The added THC was too much. They squeaked and barely scraped through their Cognitive Function testing because they were off their faces.

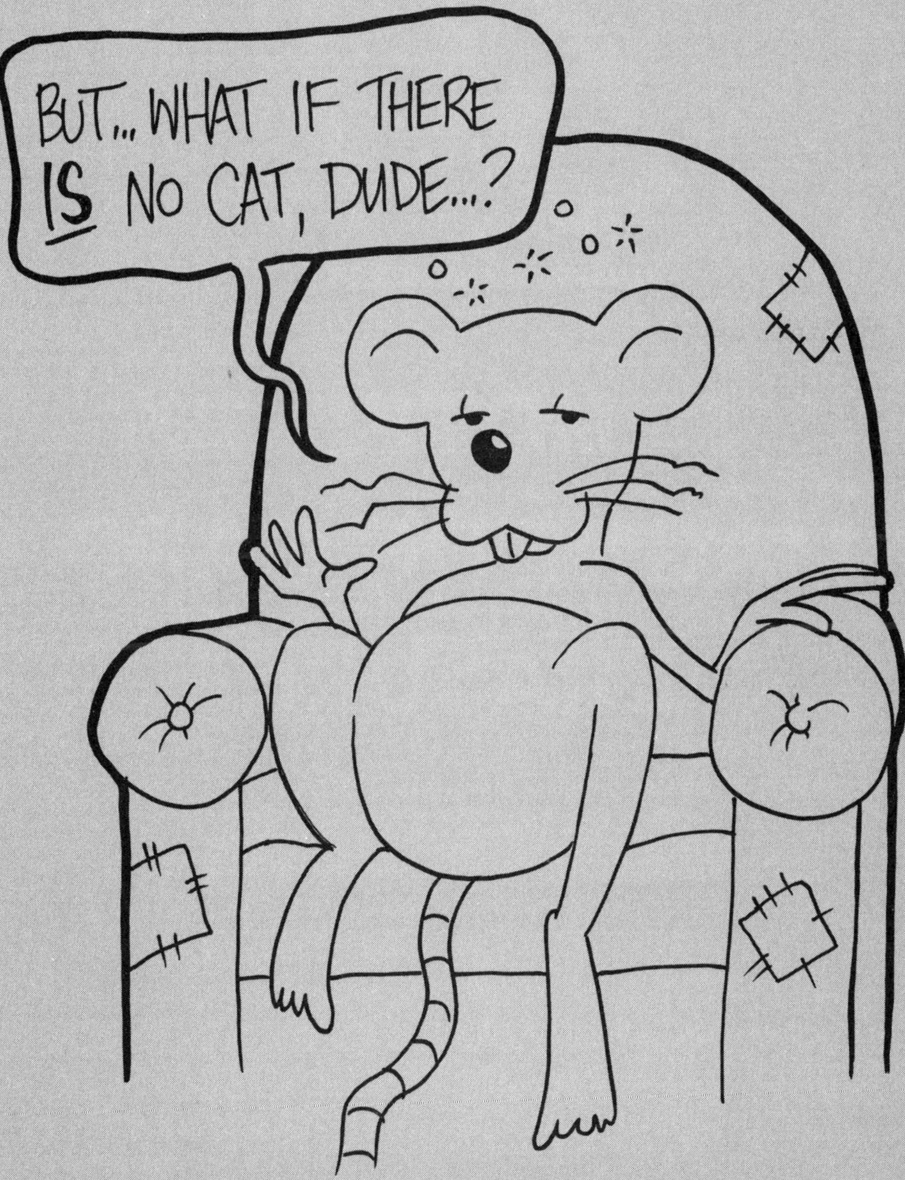
BUT... WHAT IF THERE IS NO CAT, DUDE...?

30

HOTEL MAGNETIC CARD ERASURE

IF YOU'VE STAYED in hotels recently, you've probably found that at least once, your hotel magnetic card won't open your door. No worries. You return to the front desk to get a new card. Sometimes they ask, "Have you had your hotel card near your phone?"

What's going on? Your credit cards have a magnetic stripe, too (even though you're more likely to use the chip nowadays) – and they don't get wiped by your smartphone.

Part of the reason is that your hotel magnetic card and your credit card use different magnetic materials. The other part of the reason is that your smartphone generates magnetic fields (but very weak ones). In fact, any device with electricity running through it generates magnetic fields.

MAGNETIC STRIPE 101

The magnetic stripe story begins in 1821, when Danish physicist Hans Christian Oersted ran an electric current through a wire. Suddenly, the needle on a nearby compass swung to a new position. He had discovered how to create a magnetic field.

His countryman Valdemar Poulsen took the next step in 1898. He created a magnetic field by running an electric current through a wire – just like Oersted. But he did two things differently. First, he varied the current, which varied the magnetic field. Second, he ran his now-varying magnetic field slowly along a special wire. This wire was made of a metal that could be easily magnetised. When he took away his magnetic field, the wire remained magnetised. Some parts of the wire were more magnetised, but other parts were less magnetised. He had found how to put data onto a magnetic material.

In 1928, German-Austrian engineer Fritz Pfleumer invented magnetic tape – for recording sound. Magnetic technology was becoming more mature.

In the 1960s, the London Transit Authority first used magnetic stripe technology in the busy London Underground subway system. This technology was later adopted in San Francisco for the Bay Area Rapid Transit transportation system.

Banks need a higher degree of security. So it took until around 1980 before magnetic stripes appeared on bank credit cards.

There are actually three invisible tracks on your credit card's magnetic stripe. Track 1 has a high bit density (210 bits per inch) and it contains your name, account number, expiration date, and possibly your personal identification number, etc. Track 2 has a lower density (75 bits per inch), and contains everything on Track 1 except for your name. Track 3 is usually left blank.

THE 70s HOTEL CARD - WITH 8-TRACK MAGNET

MAGNETIC STRIPE CLOSE-UP

Today, there are two main magnetic materials used.

Ferric oxide has tiny needle-like particles about 20 millionths of a metre long. These particles can be magnetised – which means they can store data. Ferric oxide magnetic stripes tend to be brown.

The other magnetic material is barium ferrite. These particles look like tiny plates. Barium ferrite magnetic stripes are usually black.

HOTEL MAGNETIC CARD ERASURE

There's a big difference between these magnetic materials.

Let me introduce you to the word "coercivity". From the sound of it, it refers to the ability to be coerced, or forced into something. But in the Land of Magnetic Physics, it refers to the ability of something to be magnetised.

Ferric oxide has a low magnetic coercivity – around 300 oersteds. (Yep, if you discover something amazing in science, you could get it named after yourself.) This tends to be used in hotel magnetic cards. It's easy to magnetise and, yes, it's easy to de-magnetise. So the weak magnetic field associated with your phone whenever it's electrically active, could wipe your hotel magnetic door card.

But barium ferrite has a much higher magnetic coercivity – up to 4000 oersteds. The advantage is that it's much less likely to be wiped by the weak magnetic field associated with your smartphone. That's why they use it in credit cards. The disadvantage is that if you want to rewrite this magnetic material, you need a much more powerful and expensive magnetic card writer. So don't carry around those small, and very powerful, rare-earth magnets. They'll easily wipe the magnetic stripe on your credit card.

So with a credit card, you can put your money where your magnet is.

31

IG NOBEL PRIZES 2016

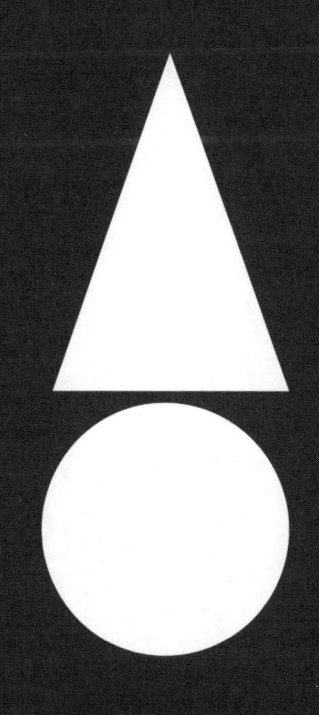

MOST OF US have heard of the Nobel Prizes, awarded for work that is surprising and original, important and deep. But not everybody has heard of the "comedy" version, the Ig Nobel Prizes.

Originally, the Ig Nobels were given for research that cannot or should not be done. They have since evolved to recognise research that makes you laugh, then think.

In 2016, Ig Nobels were awarded for research involving rats in tiny trousers, pseudoscientific gibberish, rocks with personalities – and seven other equally worthy topics.

It's always fun to run through them – so here we go.

1. BIOLOGY

The Ig Nobel Prize for Biology was awarded jointly to two men.

One winner, Thomas Thwaites, wore prosthetic arm and leg extensions so he could live in the Swiss Alps – with goats. After he successfully infiltrated their herd, he spent three days bleating, eating grass and stumbling over rocks. His discovery? Goats aren't great conversationalists. "They're much more about smell and body posture."

The other winner, Charles Foster, lived in the wild (often with his son), dressing up as various animals. These included a badger, an otter, a deer, a fox and a bird, the swift. When he lived as a badger, he ate worms, dug a den in a hillside, and tried to sniff out voles. When he became an urban fox, he "scavenged through trash and slept in gardens".

He took his task seriously. He found that worms had different flavours, depending on their location. Worms from the Chablis region of France had "a long, mineral finish", from Picardy they were "musty, like splintered wood", but from the high Kent Weald they were "fresh and uncomplicated".

2. CHEMISTRY

The Ig Nobel for Chemistry was given to the car company Volkswagen, for "solving the problem of excessive automobile pollution emissions by automatically, electro-mechanically producing fewer emissions whenever the cars are being tested". Volkswagen didn't apply for this award!

Volkswagen had installed "special" software on over half a million diesel cars. When the software detected that a car was being tested in a government laboratory, it sacrificed both fuel economy and power to reduce its emissions and "pass" the test.

But back on the road, the illegal software let the emissions rise up to 40 times the legal limit. Volkswagen was offered a $10 trillion Zimbabwean bill (worth about 40 US cents) to help with their multi-billion-dollar fines and legal costs.

Nobody from Volkswagen attended the Ig Nobel Ceremony.

3. ECONOMICS

The Economics Prize went to examining one of the foundations of Marketing – Brand Personality.

This concept defines Brand Personality as "the set of human characteristics associated with the brand". According to the theory, there would be different perceived Personalities (Sincerity, Excitement, Competence, Sophistication and Ruggedness) for a can of soup, a car or a lipstick.

But there have been niggling doubts as to the validity of Brand Personality. Dubious aspects include the use of leading questions, the fact that there are no negative values allowed – and that one of the early proponents of it, James Vicary, also invented the totally fraudulent marketing tool of Subliminal Advertising. ("Subliminal Advertising" was, and is, a total lie. The original claimed "research" was never actually carried out – it was an absolute fabrication. Furthermore, follow-up research has not been able to show that Subliminal Advertising works.)

The researchers asked several hundred students to evaluate the Personalities of several rocks. Their conclusion? "To suggest that such theory has 'no place in marketing' sounds reasonable."

So, in other words, Brand Personality is a dud.

4. LITERATURE

The Ig Nobel Prize in Literature was given to an amateur entomologist in Sweden, Fredrik Sjöberg, "for his three-volume autobiographical work about the pleasure of collecting flies that are dead, and flies that are not yet dead".

His beautifully written book *The Fly Trap* describes many things, including his fascination with the hoverfly, a master of mimicry that can copy more dangerous insects to avoid being attacked.

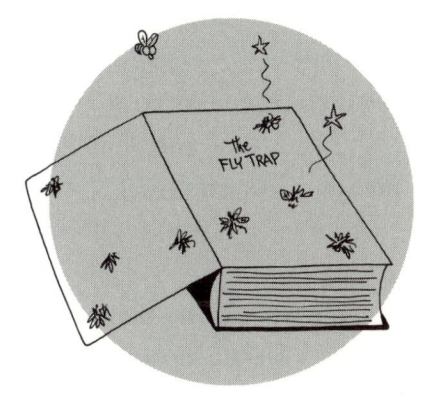

5. MEDICINE

The Ig Nobel Medicine Prize was given for research into that poorly understood phenomenon called "itch".

We all know that in many cases, scratching the itch will make it go away. And we know that a mirror tricks you into thinking your right shoulder is your left shoulder. A German team discovered that if you have an itch on the left shoulder, look into a mirror, and scratch your itch-free right shoulder, then the itch still goes away.

We don't know the reason why scratching the non-itchy side eases the itch on the itchy side.

We do know that some unexplained benefits can happen when amputees and stroke victims use mirrors in carefully planned therapy.

And if you are scratching on a non-itchy area, you are reducing the amount of scratching (and damage) on the itchy area – "sharing the love around".

6. PEACE

Some people are susceptible to believing sciencey-sounding "mumbo jumbo". So the Peace Prize was given for the paper – and there is a Language Warning – "On the Reception and Detection of Pseudo-Profound Bullshit".

For example, the sentence "Attention and intention are the mechanics of manifestation" sounds deep, but is actually meaningless. The same holds for "wholeness quiets infinite phenomena" and "hidden meaning transforms unparalleled abstract beauty". They are all just a collection of fancy nouns assembled together – and sometimes without a verb!

Supposedly, the people who fall for these kinds of meaningless phrases are "less reflective, lower in cognitive ability . . . more prone to . . . confusions and conspiratorial ideation . . . more likely to hold religious and paranormal beliefs, and are more likely to endorse complementary and alternative medicine".

I think that pretty well covers most of us . . .

7. PERCEPTION

The Perception Prize was awarded to Japanese researchers. They carried out a careful study in which some of their volunteers wore glasses that turned their vision upside down – to make it appear that they were bent over. The researchers discovered that reality appears different when you bend over and look at the world through your legs.

(And which kid doesn't know that?)

8. PHYSICS

The Physics Prize dealt with polarised light and flies. It was awarded for two separate studies, which were carried out by overlapping teams of researchers.

Horseflies are less likely to see (and bite) a white horse – because white horses emit smaller amounts of polarised light than dark horses. I guess this partly compensates for their higher rates of skin cancer, due to their white coat not blocking UV light very well.

Moving on to dragonflies, it seems they often confuse black tombstones with water, because they each emit very similar amounts of polarised light. It's often a Fatal Attraction.

So instead of a Splash, it's a Graveyard Smash!

9. PSYCHOLOGY

The Psychology Prize was awarded for detecting that the telling of lies follows an inverted U-shaped curve as we age.

The researchers looked at two measures: how frequently people told lies, and how proficient they were at telling lies. The results were very similar for each measure. Children told some lies, and were moderately good at it. Adolescents and young adults told lots and lots of lies, and were quite good at it. But as adults, they told fewer lies, and were worse at telling them.

The authors did acknowledge that the liars might have been lying to them.

10. REPRODUCTION

Finally, the Ig Nobel Prize for Reproduction was awarded for putting rats into tiny trousers, for months at a time.

If the trousers contained lots of polyester and synthetic materials, the sexual activity of the rats dropped markedly. But this drop did not happen with pants made from natural materials, such as cotton or wool.

LAUGH, THEN LEARN?

The Ig Nobel Prizes are more than just a comedic slant on Science and Medicine. They make you laugh, and then learn.

Consider the 2015 Ig Nobel Prize that was given to an Australian scientist for being able to un-boil an egg.

In other words, you could start with a raw egg, boil it to make it hard – and then apply science to make it "liquidy" and gloopy again. This sounds fun, but what's the point?

With this technology, you can easily make some very complicated chemicals. There are some pharmaceuticals that are so complicated to manufacture that they can cost thousands or tens of thousands of dollars per dose. Perhaps using this new scientific knowledge we could make these tricky medicines more cheaply, make them more widely available, and earn more dollars for Australia.

So long live the Nobel Prizes, but last laugh to the Ig Nobel Prizes.

My Ig Nobel Prize

Just in case I haven't already mentioned this, in 2002, I was awarded the Ig Nobel Prize for Interdisciplinary Research for my groundbreaking work in discovering what causes Belly Button Fluff, and why it is almost always blue! (This is in my 20th book, *Q&A with Dr K.*)

For this, I was flown from Sydney to Harvard, and accommodated, entirely at my own expense. The Ig Nobel Committee didn't want to insult me by offering me cash.

The actual prize in 2002 was a set of wind-up red chattering teeth, mounted on a stick.

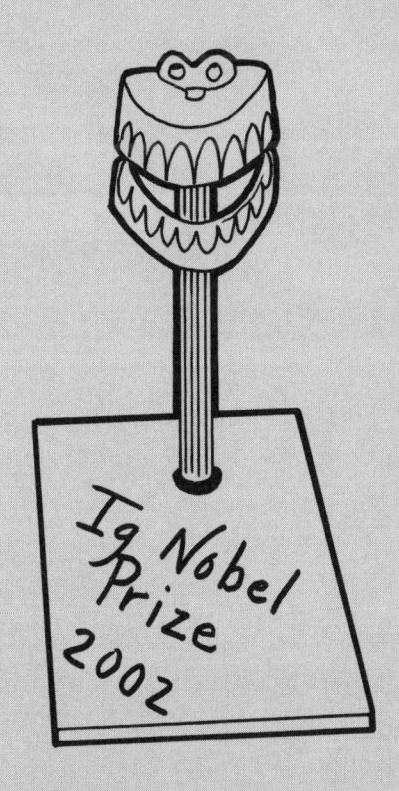

32

WIFI SPY

THE INTERNET, AND the World Wide Web, has the potential to usher us into a fabulous world of shared knowledge, social interactions, and perhaps even the redressing of disadvantages. WiFi is often the access to the Interwebs. That's the upside.

But there's always the dark side. WiFi can also spy on us. WiFi can recognise people moving about in another room – even if the WiFi unit isn't in that room. It can even work out what people are saying.

Amazingly, this new WiFi spy technology does not monitor the actual content of the WiFi radio signal – but instead, it looks at the signal's strength and timing.

How is this so?

WIFI AND FLESH

The romantically inclined might think that we humans are beautifully evolved mobile beings of physical grace and enlightened consciousness.

But as far as a WiFi unit is concerned, we are just a mobile bag of meat, gristle and water – that gets in the way of its signals. Our bodies (whether moving or stationary) interfere with the signals that the WiFi unit sends out and receives.

By analysing the strength and timing of the signal returning to the WiFi unit, you can work out if people are about, what movements their bodies, arms and legs are making – and even what sounds are coming from their mouths.

The folk who live in cities spend virtually each second of their day drenched in WiFi. (By the way, it has never been proved that WiFi has harmful health effects on humans.)

Given we are constantly bathing in WiFi, it's helpful to know some of its history – which now goes back nearly half a century.

What Does the "Fi" in Wi-Fi Mean?

Nothing.

Back in 1999, there was in existence a rather interesting technology called "IEEE 802.11b Direct Sequence". But this name wasn't very catchy. So the brand-consulting firm Interbrand Corporation was hired to come up with something better.

Now, a few decades prior to this, a home audio system had only a record player, amplifier and speakers. Mostly, the quality of the home system was not very good. But the high-end version did exist and was called High Fidelity. Pretty soon, this got shortened to Hi-Fi.

By the time Interbrand were hired, the term HiFi had been degraded to describe any consumer-grade audio equipment – and most people understood its new meaning. So Interbrand "leaned" on this, and created the nonsense name "WiFi" – where "Wi" stood for wireless, and "Fi" stood for nothing at all. It was derivative and catchy – and it rhymed!

In an effort to make sense of the "Fi", some people have suggested that it stands for "frequency interface". This is because anything that adapts to, or talks to, a network is classified as an "interface".

But in reality, the "Fi" in WiFi stands for nothing – but it's a whole lot easier to remember and say than "IEEE 802.11b".

WIFI HISTORY

The basic concept of WiFi is that we can send data by electromagnetic waves. Today, WiFi operates at microwave frequencies – but at very low power. This is why the more powerful microwave cooker in your kitchen can interfere with your much weaker WiFi unit. (See "Microwave Oven Hertz Your WiFi" in my 32nd book, *50 Shades of Grey Matter*.)

One of the earliest versions of WiFi was ALOHAnet in 1971, which connected the Hawaiian islands. But ALOHAnet transmitted and received at radio frequencies. It used much lower frequencies than we use today – it ran at millions of hertz, rather than billions.

In 1991, the NCR Corporation and the telco AT&T invented another variety of WiFi. They used it to link cash registers and cashier systems.

A team of Australian CSIRO scientists, led by Dr John O'Sullivan, provided an essential part of modern WiFi. It happened accidentally when they carried out what Dr O'Sullivan described as "a failed experiment to detect exploding mini black holes the size of [atoms]". The CSIRO patents that mathematically allowed the WiFi signal to be "unsmeared" were granted in 1992 and 1996. Since then, these patents have earned over a billion dollars in royalties. (See "WiFi and Black Holes" in my 32nd book, *50 Shades of Grey Matter*.)

I first started using WiFi in the year 2000. Back then it was pretty slow and clunky – especially with dial-up internet! However, since then, WiFi has become almost as reliable and easy to use as a light switch. Your smartphones, computers (desktop or laptop), tablets and smartwatches each carry tiny WiFi units inside.

WIFI 101

A WiFi unit (or router) is basically a radio transmitter and receiver. But unlike old-fashioned radio, which carried analogue voice and music, WiFi carries digital data.

It allows access to the internet for all your devices. Today's WiFi mostly works at frequencies of 2.4 and 5 gigahertz – which corresponds to wavelengths of 12 and 6 centimetres. Those wavelengths are similar to dimensions that exist in the human body. (Your hand is about 10 centimetres across.) So WiFi signals are affected by objects around that size.

A WiFi router will receive a request for information – for example, today's weather forecast – then broadcast a signal that answers that request, and finally receive the reflected echo. The WiFi router analyses that reflected echo. It needs to know how well its signals travel into, and back from, its local environment, so that it can adjust its outgoing signal for more reliable communication. The size, shape and timing of the reflected echo depends on what's in the environment and where the echoes are coming from. This information (size, etc.) is called Channel State Information. Channel State Information is what the new WiFi spy technology is analysing.

WIFI SPY – GOVERNMENT

By a kind of coincidence, WiFi frequencies are close to those used in radar.

Indeed, the radar community had been experimenting with "looking" through walls, to rescue people trapped by a fire or a collapsing building, or to warn law enforcement officers of an ambush. But until around 2010, this was high-level military grade gear, requiring a few gigahertz of bandwidth (huge), a 2.4-metre-long antenna (too large to carry easily) and a huge power supply.

Only the military or law enforcement agencies could easily use this gear.

WIFI SPY – DOMESTIC

However, by 2012, the technology had advanced significantly, and the available equipment had changed dramatically. Engineers worked out how to look through a wall with regular domestic WiFi equipment that used only a few tens of megahertz of bandwidth – and was small enough to hold in your hand.

Back in 2012, they needed two transmitting units, one on each side of the wall. They could "look" through a 15-centimetre hollow wall and reliably tell how many moving people were in a closed room – but only if the number of people was three or fewer. (At the time, four or more people "blurred" the reflected WiFi signals too much for the technology to pick them apart.)

In the case of a single person sending messages based on gesture (e.g. push, pull, punch), they could decode all messages reliably if the person were closer to the WiFi unit than 5 metres. However, the results were much worse when analysing groups of people.

After all, a human is not a single object, but a bunch of different body parts, all of them moving in a loosely coupled way. For example, when you are walking, your torso or trunk might be moving at one metre per second, while your arms and legs might be swinging back and forth at two metres per second. We each have our own slightly different walking styles.

But this "difference between people" that initially made the results harder to interpret turned out to have a benefit.

The engineers could now identify individuals by those differences.

HOW THE WIFI SPY CAPTURES YOUR FIGURE

CAPTURED FIGURE

NO PLACE TO HIDE

The WiFi Spy technology has advanced to the stage where it can pick differences between up to half a dozen people standing in a room.

One test early in 2016 achieved human identification accuracy of 77 per cent with six people in the group, and 93 per cent with just two people. Another test with improved technology in late 2016 did better again with 89 per cent and 94 per cent accuracy.

In terms of detecting and identifying smaller body parts and their movements, a WiFi system called "WiKey" can work out what keys your fingers are tapping on a keyboard. After some training, WiKey reached 93.5 per cent accuracy. Remember, this is done by looking only at the strength, time, etc. of the reflected WiFi signal as it bounces back into the router. It does not look at the content of the WiFi signal.

A similar WiFi system can detect what movements (and therefore sounds) your mouth is making. After all, your mouth makes different shapes to enunciate different sounds. This WiFi spy can pick the difference between your various mouth movements. By 2016, the recognition accuracy was 91 per cent for a single user, and up to 74 per cent for three people speaking simultaneously.

Mind you, some of these test systems used modified WiFi routers, and often at very close range of a metre or so. But overall, the technology is geared to seeing what can be done with Commercial Off-The-Shelf systems. Perhaps that's so this spy technology will work with your average home WiFi?

So mentally prepare for a world where your WiFi can switch on your rice cooker before you come home, let the dog out, or call for help in an emergency. But be aware that with the WiFi spy looking in, we could be the last generation to have ever known privacy.

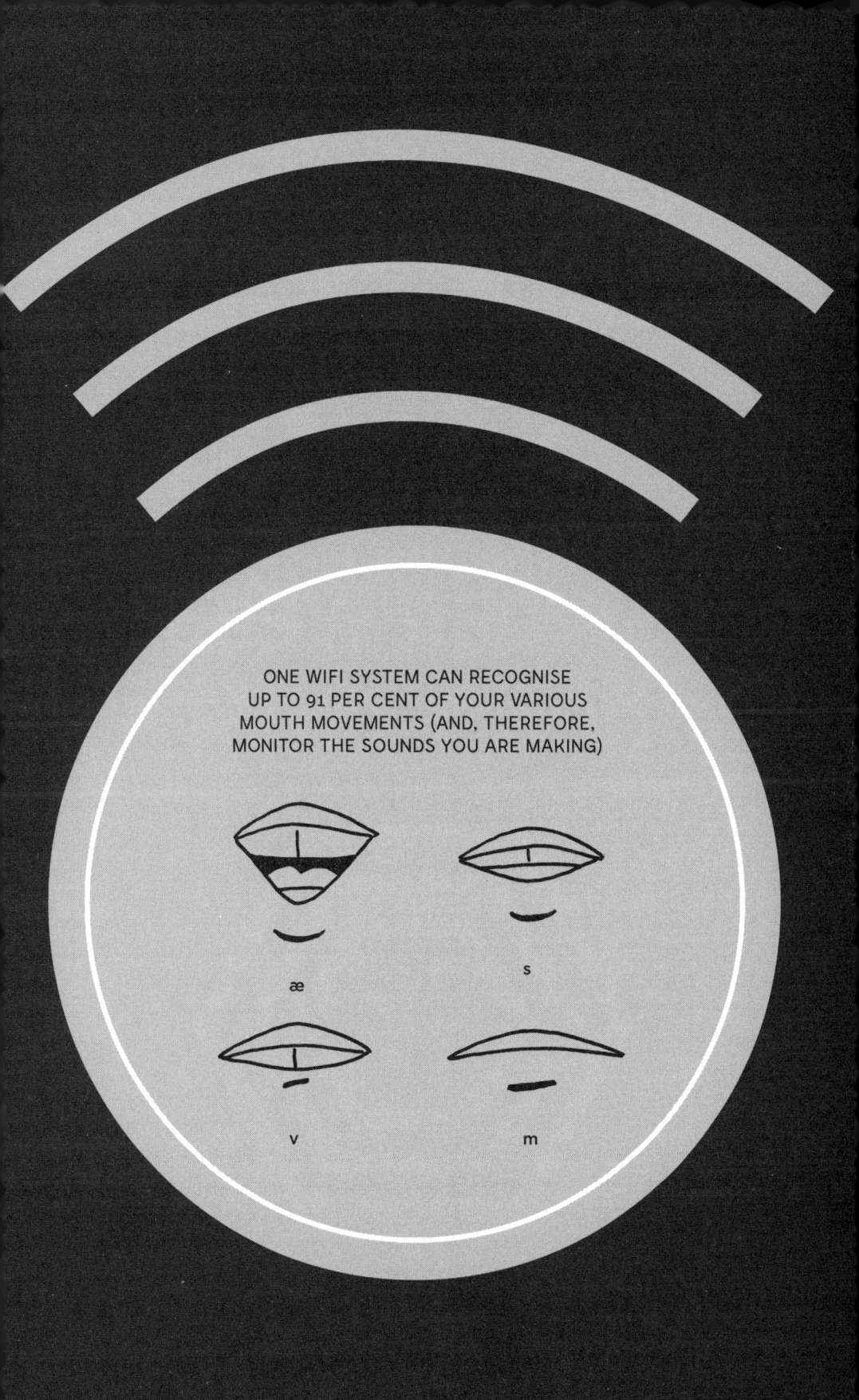

ONE WIFI SYSTEM CAN RECOGNISE UP TO 91 PER CENT OF YOUR VARIOUS MOUTH MOVEMENTS (AND, THEREFORE, MONITOR THE SOUNDS YOU ARE MAKING)

æ

s

v

m

33
SPIDER WEBS DON'T TWIST

IF YOU'VE EVER seen a spider abseiling down its dragline silk, you might not have noticed something very important – because it wasn't there to see. As the spider gracefully descended down its silk, you did not see it twisting. But when human commandos or Special Forces Operatives abseil down a rope from a helicopter, they do twist.

How come a spider suspended on its silk thread does not twist in the wind?

It's something the more curious of us have marvelled at for thousands of years – and only recently did we find the answer. The spider turns out to be "immune" to twisting because of special proteins inside the web thread.

VERY EASY TO TWIST

Surprisingly, it takes very little force to make a hanging wire twist.

In 1778, British physicist Henry Cavendish used this little oddity of Nature to measure the mass of the Earth.

He suspended heavy and light lead balls (158 kilograms and 0.71 kilograms) on wires. Because they had mass, they had gravity – and so they attracted each other. (Mind you, the attraction was very, very small.) He cleverly arranged them in such a way that when they did attract each other, the suspension wires would twist. Of course, the amount of "twist" was truly microscopic. He had to measure this tiny amount of twist.

After two years of painstaking and delicate work, he was able to work out the mass of the entire Earth – and he got it accurate to within 1 per cent. The precision of this experiment was absolutely amazing – and over two centuries ago, he had no computers or other high-powered technology to help him!

But the point is this. If hanging wires twist so easily in response to external forces, what special "stuff" is there inside spider silk that stops it twisting?

SENSATIONAL SPIDER SILK

Weight for weight, spider silk is stronger than high-strength steel. Yet, unlike steel, it is totally made from organic chemicals – and is manufactured at room temperature.

Spider silk also has other remarkable mechanical properties – such as its ability to either contract or extend. It also has an exceptional ability to carry heat. But only recently have we begun to explore its ability to twist – and to untwist.

When spider silk is twisted and let go, two things happen.

First, very quickly, the oscillations die away. It stops twisting. But this is the opposite of what happens if you hang a weight from the end of a fishing line, twist it around, and then release it. In that situation, the fishing line will keep on twisting clockwise and anticlockwise for ages – oscillating back and forth.

Second, when the oscillations of the spider silk come to a halt, the spider silk has permanently twisted around to have a new rest point. Suppose that you put a tiny red dot on the spider silk, and that the dot is pointing to 12 o'clock. Then twist the silk around clockwise to 5 o'clock and let go. It will very quickly stop twisting – but now the red dot has been reset to 2 o'clock.

So, this is a big part of why spider silk stops twisting so quickly. About 75 per cent of the energy in the oscillations has been robbed – and put into permanently deforming the spider silk. So spider silk is not totally elastic.

One astonishing thing about spider silk is that you can do this one, 10 or 100 times – and the spider silk will keep resetting, but will not break. So the spider silk simultaneously has a degree of elasticity and of non-elasticity.

ELASTIC VERSUS PLASTIC

"Elastic" means that after you deform something, it will return to its original state.

Rubber bands are elastic. So is a tennis ball. If you push a dent into a tennis ball with your thumbs, and then let go, the tennis ball recovers its original shape.

"Plastic" means that if you deform something and then let go, it doesn't bounce back. It stays where you pushed it to. Plasticine is plastic. So is wood.

Of course, no material is 100 per cent elastic or plastic.

Elastic rubber bands have a tiny amount of plasticity. After severe stretching, they return almost all the way back to their original shape – but not all the way.

Wood is not entirely plastic. If you put a dent into wood, it will recover a tiny amount.

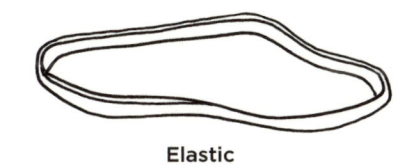

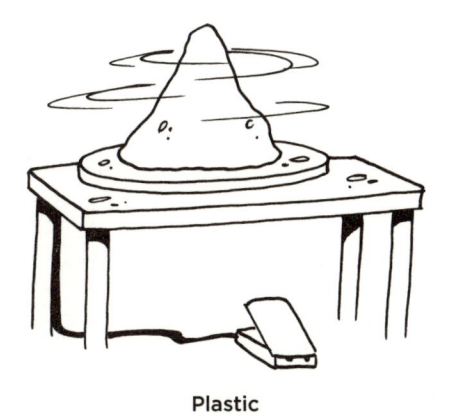

Elastic

Plastic

Mr Fantastic

SPIDER SILK 101

A single strand of spider silk has an outer "skin". Inside this outer skin is a solid cylinder that is pierced by many fibres (called "fibrils").

SPIDER SILK

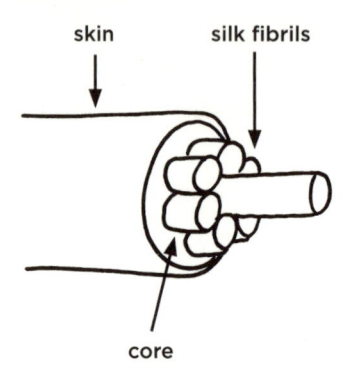

skin — silk fibrils

core

The fibrils are in turn made up of two types of proteins.

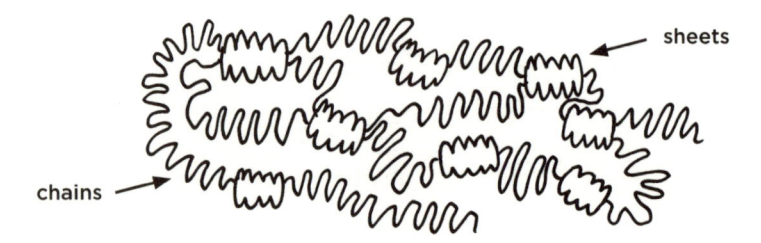

sheets

chains

First, there are "chains" of proteins. These are mostly plastic: when they are pulled, they permanently change shape – and it takes energy to do this. This is why the oscillations die down so quickly, and why the spider doesn't twist around as it sails down its dragline silk. The result is that the spider is less likely to attract attention to itself – and also is better able to keep an eye on the outside world for both predators and food.

"CHAINS" BEING PULLED

Second, there are "sheets" of proteins, which are folded on top of each other many times. (They look like slabs of 2-Minute Noodles.) They are mostly elastic. This means that when they are pulled, they return to their original shape.

"SHEETS" BEING PULLED

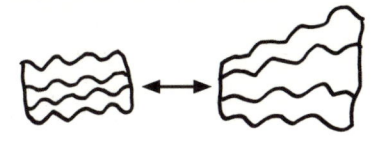

THE FUTURE

So now that we are beginning to understand why the suspended spider doesn't twist around, we might be able to apply this knowledge to human technology.

We could have guitar and violin strings that don't twist, as well as parachute cords, and ropes for mountain climbers and window cleaners.

And perhaps most importantly of all, headphone cables that don't get twisted into an unfathomable knot!

REFERENCES

01 SEXIST AIR-CON

"Building Emissions: Female Thermal Demand", by Joost van Hoof, *Nature Climate Change*, December 2015, Vol. 5, No. 12, pages 1029–1030.

"Chilly At Work? Office Formula Was Devised for Men", by Pam Belluck, *The New York Times*, 3 August 2015, https://www.nytimes.com/2015/08/04/science/chilly-at-work-a-decades-old-formula-may-be-to-blame.html

"Energy Consumption in Buildings and Female Thermal Demand", by Boris Kingma and Wouter van Marken Lichtenbelt, *Nature Climate Change*, December 2015, Vol. 5, No. 12, pages 1054–1056.

"Is Your Thermostat Sexist?", by Anthony Lydgate, *The New Yorker*, 3 August 2015.

"New Technologies Aim to Save Energy – and Lives – with Better Air-Conditioning", by David Biello, *Scientific American*, 6 September 2013, https://www.scientificamerican.com/article/technology-improvements-save-energy-and-lives-with-air-conditioning/

"Why Many Women Shiver in the Office", by Robyn Williams and Wendy Zukerman, *The Science Show*, 15 August 2015, http://www.abc.net.au/radionational/programs/scienceshow/why-many-women-shiver-in-the-office/6697914#transcript

"Why Men and Women Battle over the Office Thermostat", by Nick Stockton, *Wired*, 3 August 2015, https://www.wired.com/2015/08/men-women-battle-office-thermostat/

"Women Shiver at Work in 'Sexist' Air Conditioning", by Sarah Knapton, *The Telegraph*, 3 August 2015, http://www.telegraph.co.uk/science/2016/03/15/women-shiver-at-work-in-sexist-air-conditioning/

02 COFFEE NAP

"Caffeine Disposition after Oral Doses", by Maurizio Bonati et al., *Clinical Pharmacology and Therapeutics*, July 1982, Vol. 32, No. 1, pages 98–106.

"Health Check: What Are 'Coffee Naps' and Can They Help You Power Through the Day?", by Chin Moi Chow, *The Conversation*, 3 April 2017, https://theconversation.com/health-check-what-are-coffee-naps-and-can-they-help-you-power-through-the-day-73952

"Suppression of Sleepiness in Drivers: Combination of Caffeine with a Short Nap", by Louise A. Reyner and James A. Horne, *Psychophysiology*, November 1997, Vol. 34, No. 6, pages 721–725.

03 POO IN 12 SECONDS

"All Mammals Poop in 12 Seconds and There's an Equation for the 'Duration of Diarrheal Defecation'", by Sara Chodosh, *Flipboard*, 28 April 2017, https://flipboard.com/@flipboard/flip.it%2F3U_tBF-all-mammals-poop-in-12-seconds-and-ther/f-a6a5f729e2%2Fpopsci.com

"Hydrodynamics of Defecation", by P.J. Yang et al., *Soft Matter*, 25 April 2017, DOI:10.1039/C6SM02795D

"Physics of Poo: Why It Takes You and An Elephant the Same Amount of Time", by David Hu and Patricia Yang, *The Conversation*, 27 April 2017, http://theconversation.com/physics-of-poo-why-it-takes-you-and-an-elephant-the-same-amount-of-time-76696

04 BRAVE NEW WOMB

"An Extra-Uterine System to Physiologically Support the Extreme Premature Lamb", by Emily A. Partridge et al., *Nature Communications*, 25 April 2017, doi:10.1038/ncomms15112

"Brave New Wool? Artificial Womb Sustains Premature Lambs for Weeks", by Ike Swetlitz, *Scientific American*, 25 April 2017, https://www.scientificamerican.com/article/brave-new-wool-artificial-womb-sustains-premature-lambs-for-weeks

REFERENCES

"Faux Womb Keeps Preemie Lambs Alive", by Tina Hesman Saey, *ScienceNews*, 25 April 2017, https://www.sciencenews.org/article/faux-womb-keeps-preemie-lambs-alive

"Premature Lambs Kept Alive in 'Plastic Bag' Womb", by Michelle Roberts, *BBC News*, 25 April 2017, http://www.bbc.com/news/health-39693851

"Preterm Birth: Fact Sheet", World Health Organization, November 2016, http://www.who.int/mediacentre/factsheets/fs363/en

05 EARWORM

"Can't Get Kylie Out of Your Head? Blame the Shape of Your Brain", by Sam Wong, *New Scientist*, 7 July 2015, https://www.newscientist.com/article/dn27919-cant-get-kylie-out-of-your-head-blame-the-shape-of-your-brain

"Can't Get You Out of My Head – What Makes a Song an Earworm", by Nicola Davis, *The Guardian*, 4 November 2016, https://www.theguardian.com/science/2016/nov/03/cant-get-you-out-of-my-head-what-makes-a-song-an-earworm-lady-gaga

"Dissecting an Earworm: Melodic Features and Song Popularity Predict Involuntary Musical Imagery", by Kelly Jakubowski et al., *Psychology of Aesthetics, Creativity and the Arts*, May 2017, Vol. 11, No. 2, pages 122–135.

"How Chewing Gum Can Stop Unwanted Songs from Playing in Your Head", by David DiSalvo, *Forbes*, 10 May 2015, https://www.forbes.com/sites/daviddisalvo/2015/05/10/how-chewing-gum-can-stop-unwanted-songs-from-playing-in-your-head

"Psychologists Identify Key Characteristics of Earworms", 3 November 2016, http://www.apa.org/news/press/releases/2016/11/earworms.aspx

"Tunes Stuck in Your Brain: The Frequency and Affective Evaluation of Involuntary Musical Imagery Correlate with Cortical Structure", by Nicolas Farrugia et al., *Consciousness and Cognition*, September 2015, Vol. 35, pages 66–77.

"Want to Block Earworms from Conscious Awareness? B(u)y Gum!", by

C. Philip Beaman et al., *The Quarterly Journal of Experimental Psychology*, 21 April 2015, Vol. 68, No. 6, DOI:10.1080/17470218.2015.1034142

"What Lady Gaga's 'Bad Romance' and Other Earworm Songs Have in Common", by Joanna Klein, *The New York Times*, 3 November 2016, https://www.nytimes.com/2016/11/04/science/earworms-lady-gaga-bad-romance.html

06 HYDROTHERMAL VENTS AND INVISIBLE MOUNTAIN RANGE

"Ongoing Hydrothermal Activities within Enceladus", by Hsiang-Wen Hsu et al., *Nature*, 12 March 2015, Vol. 519, No. 7542, pages 207–210.

"The Physiology and Habitat of the Last Universal Common Ancestor", by Madeleine C. Weiss et al., *Nature Microbiology*, September 2016, DOI: 10.1038/nmicrobiol.2016.116

"We've Been Wrong about the Origins of Life for 90 Years", by Arunas L. Radzvilavicius, *Quartz*, 30 August 2016, https://qz.com/761430/weve-been-wrong-about-the-origins-of-life-for-90-years

07 LIFE ON ENCELADUS

"Buried 'Lake Superior' Seen on Saturn's Moon Enceladus", by Lisa Grossman, *New Scientist*, 3 April 2014, https://www.newscientist.com/article/dn25355-buried-lake-superior-seen-on-saturns-moon-enceladus

"Cassini Embarks on Twilight Mission at Saturn", by Paul Voosen, *Science*, 14 April 2017, Vol. 356, No. 6334, pages 120–121.

"Cassini Finds Molecular Hydrogen in the Enceladus Plume: Evidence for Hydrothermal Processes", by J. Hunter Waite et. al., *Science*, 14 April 2017, Vol. 356, No. 6334, pages 155–159.

"Cassini Nears the End of a 13-Year Saturn Exploration", by Frank Morring Jr, *Aviation Week and Space Technology*, 3 April 2017, page 18.

"Cassini Plumbs the Depths of the Enceladus Sea", by Richard Kerr, *Science*, 4 April 2014, Vol. 344, No. 6179, page 17.

REFERENCES

"Cassini's Science Swan-Song" by Alexandra Witze, *Nature*, 12 April 2017, Vol. 544, No. 7649, pages 149–150.

"Detecting Molecular Hydrogen on Enceladus", by Jeffrey S. Seewald, *Science*, 14 April 2017, Vol. 356, No. 6334, pages 132–133.

"Enceladus' Hot Springs", by Gabriel Tobie, *Nature*, 12 March 2015, Vol. 519, No. 7542, pages 162–163.

"Follow the Plume: The Habitability of Enceladus", by Christopher P. McKay et al., *Astrobiology*, April 2014, Vol. 14, No. 4, pages 352–355.

"Food for Microbes Found on Enceladus", by Ashley Yeager, *Science News*, 13 April 2017, https://www.sciencenews.org/article/food-microbes-found-enceladus

"Global Ocean Extends across Entire Core of Saturn's Moon, Enceladus, NASA confirms", *ABC News*, 16 September 2015, http://www.abc.net.au/news/2015–09–16/nasa-mission-finds-global-ocean-on-saturn-moon/6780456

"Keeping Enceladus Warm", by B.J. Travis and G. Schubert, *Icarus*, April 2015, Vol. 250, pages 32–42.

"Liquid Ocean Sloshes under Saturn's Moon's Icy Crust, Cassini Evidence Shows", by Clara Moskowitz, *Scientific American*, 3 April 2014, https://www.scientificamerican.com/article/liquid-ocean-saturn-moon-enceladus

"NASA's Cassini Spacecraft Plunges through Saturn Moon Enceladus' Icy Spray", *ABC News*, 30 October 2015, http://www.abc.net.au/news/2015–10–30/nasa-spacecraft-plunges-through-saturn-moons-icy-spray/6897866

"Ocean Discovered on Enceladus May Be the Best Place to Look for Alien Life", by Ian Sample, *The Guardian*, 4 April 2014, https://www.theguardian.com/science/2014/apr/03/ocean-enceladus-alien-life-water-saturn-moon

"Ongoing Hydrothermal Activities within Enceladus", by Hsiang-Wen Hsu et al., *Nature*, 12 March 2015, Vol. 519, No. 7542, pages 207–210.

"The Gravity Field and Interior Structure of Enceladus", by L. Iess et al., *Science*, 4 April 2014, Vol. 344, No. 6179, pages 78–80.

"Tidal Heating in Enceladus", by Jennifer Meyer and Jack Wisdom, *Icarus*,

June 2007, Vol. 188, No. 2, pages 535–539.

"Tidally-Induced Melting Events as the Origin of South-Pole Activity on Enceladus", by Marie Běhounková et al., *Icarus*, June 2012, Vol. 219, No. 2, pages 655–664.

"The Drive to Life on Wet and Icy Worlds", by Michael J. Russell et al., *Astrobiology*, April 2014, Vol. 14, No. 4, pages 308–343.

"Under Icy Surface of a Saturn Moon Lies a Sea of Water, Scientists Say", by Kenneth Chang, *The New York Times*, 3 April 2014, https://www.nytimes.com/2014/04/04/science/space/a-moon-of-saturn-has-a-sea-scientists-say.html

"Under the Sea of Enceladus", by Frank Postberg et. al., *Scientific American*, October 2016, Vol. 315, No. 4, pages 28–35.

08 REFRIED BEANS

"What Are 'Refried Beans'?", by David Mikkelson, *Snopes*, 11 April 2017, http://www.snopes.com/food/prepare/refriedbeans.asp

09 MISOPHONIA – HATING SOUND

"Decreased Sound Tolerance and Tinnitus Retraining Therapy (TRT)", by Margaret M. Jastreboff and Pawel J. Jastreboff, *The Australian and New Zealand Journal of Audiology*, November 2002, Vol. 24, No. 2, pages 74–84.

"Etiology, Composition, Development and Maintenance of Misophonia: A Conditioned Aversive Reflex Disorder", by Thomas H. Dozier, *Psychological Thought*, 2015, Vol. 8, No. 1, pages 114–129.

"Mastication Rage: A Review of Misophonia – An Under-Recognised Symptom of Psychiatric Relevance?", by George Bruxner, *Australasian Psychiatry*, April 2016, Vol. 24, No. 2, pages 195–197.

"Misophonia: Diagnostic Criteria for a New Psychiatric Disorder", by Arjan Schröder et al., *PLOS One*, January 2012, Vol. 8, No. 1, e54706.

"Misophonia: Current Perspectives", by Andrea E. Cavanna and Stefano Seri, *Neuropsychiatric Disease and Treatment*, 2015, Vol. 11, pages 2117–2123.

REFERENCES

"Please Stop Making That Noise" by Barron H. Lerner, *The New York Times* Well blog, 23 February 2015, https://well.blogs.nytimes.com/2015/02/23/please-stop-making-that-noise

"Snip, Snap, Slurp: Misophonia Makes Them Unbearable", by Karen Barrow, *The New York Times* Well blog, March 2, 2015, https://well.blogs.nytimes.com/2015/03/02/snip-snap-slurp-misophonia-makes-them-unbearable

"The Brain Basis for Misophonia", by Sukhbinder Kumar et al., *Current Biology*, 20 February 2017, Vol. 27, No. 4, pages 527–533.

"When a Chomp or a Slurp Is a Trigger for Outrage", by Joyce Cohen, *The New York Times*, 5 September 2011, http://www.nytimes.com/2011/09/06/health/06annoy.html

"Why the Sound of Noisy Eating Fills Some People with Rage", by Tiffany O'Callaghan, *New Scientist*, 2 February 2017, https://www.newscientist.com/article/2120167-why-the-sound-of-noisy-eating-fills-some-people-with-rage/

10 SHOUTING AT HARD DRIVES

"A Loud Sound Just Shut Down a Bank's Data Center for 10 Hours", by Andrada Fiscutean, *Vice Motherboard* blog, 12 September 2016, https://motherboard.vice.com/en_us/article/8q8dqg/a-loud-sound-just-shut-down-a-banks-data-center-for-10-hours

"End User FAQs – INERGEN Clean Agent Fire Suppression System", Tyco Fire Products, 2013, https://www.ansul.com/en/us/DocMedia/F-2012095.pdf

"Fire Drill Knocks ING Bank's Data Centre Offline", *BBC News*, 12 September 2016, http://www.bbc.com/news/technology-37337868

"Inert Gas Data Center Fire Protection and Hard Disk Drive Damage", by Brian P. Rawson and Kent C. Green, *Data Center Journal*, 27 August 2012, http://www.datacenterjournal.com/inert-gas-data-center-fire-protection-and-hard-disk-drive-damage

"Mass Drive Failings Linked to Accidental Gas (Fire Suppression) Release in Sydney Data Centre", *vpsBoard* forum, 13 January 2014, https://vpsboard.com/threads/mass-drive-failings-linked-to-accidental-gas-fire-suppression-release-in-sydney-data-centre.3193

"Potential Problems with Computer Hard Discs when Fire Extinguishing Systems Are Released", Siemens Switzerland Ltd, 2010, http://www.hgi-fire.com/wp-content/uploads/2013/03/White-Paper-potential-problems-with-computer-hard-disks-V1.3.pdf

"Silent Extinguishing", Siemens AG, September 2015, http://www.buildingtechnologies.siemens.com/bt/global/en/firesafety/extinguishing/about-sinorix/latest-technical-findings/Documents/White_Paper_Extinguishing_final_en.pdf

11 PONYTAIL SWING

"Ig Nobels Go to Studies of Swaying Ponytails, (Nearly) Exploding Colons", by John Bohannon, *Science*, 21 September 2012, http://www.sciencemag.org/news/2012/09/ig-nobels-go-studies-swaying-ponytails-nearly-exploding-colons

"Joseph B. Keller, Mathematician with Whimsical Curiosity, Dies at 93", by Sam Roberts, *The New York Times*, 16 September 2016, https://www.nytimes.com/2016/09/17/us/joseph-b-keller-mathematician-with-whimsical-curiosity-dies-at-93.html

"Ponytail Motion", by Joseph Keller, *Society for Industrial and Applied Mathematics Journal on Applied Mathematics*, 2010, Vol. 70, No. 7, pages 2667–2672.

"Ponytail Physics: How Competing Forces Shape Bundles of Hair", by John Matson, *Scientific American*, 3 March 2012, https://blogs.scientificamerican.com/observations/ponytail-physics-how-competing-forces-shape-bundles-of-hair

"Science Behind Ponytail Revealed", *BBC Science & Environment*, 13 February 2012, http://www.bbc.com/news/science-environment-17012795

"Shape of a Ponytail and the Statistical Physics of Hair Fiber Bundles", by Raymond E. Goldstein et al., *Physical Review Letters*, 17 February 2012, Vol. 108, No. 7, pages 078101–1 to 078101–4.

12 LIGHTNING POWER

"A Place Where Lightning Strikes Almost 300 Days a Year", by Joanna Klein, *The New York Times*, 16 May 2016, https://www.nytimes.com/2016/05/17/science/in-venezuela-the-lightning-capital-of-the-world.html

"Bolt from the Blue: Lightning Doesn't Form Like We Thought", by Shannon Hall, *New Scientist*, 11 April 2017, https://www.newscientist.com/article/mg23431210-500-electric-atmosphere-how-lightning-is-born-shocking-discovery-rewriting-the-story-of-lightning/

"Earth's New Lightning Capital Revealed", by Molly Porter, NASA, 3 May 2016, https://www.nasa.gov/centers/marshall/news/news/releases/2016/earths-new-lightning-capital-revealed.html

"Electricity from Lightning", by Marnie Chesterton, *BBC World Service*, 5 November 2016, http://www.bbc.co.uk/programmes/p04dhs91

"Lightning, Weather's Byproduct, May Become One of Its Predictors", by Kate Murphy, 9 January 2017, *The New York Times*, https://www.nytimes.com/2017/01/09/science/lightning-weather-prediction.html

"Where Are the Lightning Hotspots on Earth?", by Rachel I. Albrecht et al., *Bulletin of the American Meteorological Society*, November 2016, pages 2051–2068.

13 TRUCK SHIFTS DATA FASTER THAN OPTIC FIBRE

"Amazon Reveals AWS Snowmobile, a 45-Foot Semi-Trailer that Moves Exabytes of Data to the Cloud", by Dan Richman, *GeekWire*, 30 November 2016, https://www.geekwire.com/2016/use-amazons-snowball-snowballs-unleashes-45-foot-truck-model

"Amazon's Snowmobile is Actually a Truck Hauling a Huge Hard Drive", by Klint Finley, 2 December 2016, *Wired*, https://www.wired.com/2016/12/amazons-snowmobile-actually-truck-hauling-huge-hard-drive/

"Amazon's Snowmobile Will Let You Upload Stuff by the Truckload – Literally", by Alex Hern, *The Guardian*, 5 December 2016, https://www.theguardian.com/technology/2016/dec/05/amazon-snowmobile-upload-truckload

"Amazon Will Truck Your Massive Piles of Data to the Cloud with an 18-Wheeler", by John Mannes, *TechCrunch*, 20 November 2016, https://techcrunch.com/2016/11/30/amazon-will-truck-your-massive-piles-of-data-to-the-cloud-with-an-18-wheeler

"AWS Snowmobile", *Amazon Web Services*, https://aws.amazon.com/snowmobile

"This Company is Using Amazon Snowmobile to Transfer Petabytes of Data to the Cloud", by Brandon Butler, *Network World*, January 5, 2017, http://www.networkworld.com/article/3154898/cloud-computing/this-company-is-using-amazon-snowmobile-to-transfer-petabytes-of-data-to-the-cloud.html

"This 'Snowmobile' Truck Seems Like a Joke, but It's Actually Amazon's Newest Product", by Julie Bort, *Business Insider Australia*, 1 December 2016, https://www.businessinsider.com.au/amazon-snowmobile-truck-is-a-funny-for-real-product-2016–11

14 THE SMELL OF BOOKS

"Can You Judge A Book by Its Odour?", by Claire Armitstead, *The Guardian*, 8 April 2017, https://www.theguardian.com/books/2017/apr/07/the-smell-of-old-books-science-libraries

"Characterisation of Compounds Emitted During Natural and Artificial Ageing of a Book. Use of Headspace-Solid-Phase Microextraction/Gas Chromatography/Mass Spectrometry", by Agnès Lattuati-Derieux et al., *Journal of Cultural Heritage*, April 2006, Vol. 7, No. 2, pages 123–133.

"Degradation Product Emission from Historic and Modern Books by Headspace SPME/GC-MS: Evaluation of Lipid Oxidation and Cellulose

Hydrolysis", by Andrew J. Clark et al., *Analytical and Bioanalytical Chemistry*, April 2011, Vol. 399, No. 10, pages 3589–3600.

"Material Degradomics: On the Smell of Old Books", by Matija Strlič et al., *Analytical Chemistry*, January 2009, Vol. 81, No. 20, pages 8617–8622.

"Smell of Heritage: A Framework for the Identification, Analysis and Archival of Historic Odours", by Cecillia Bembibre and Matilja Strlič, *Heritage Science*, 7 April 2017, Vol. 5, No. 2,, DOI:10.1186/s40494-016-0114-1

"What Causes the Smell of New & Old Books?", *Compound Interest*, 1 June 2014, http://www.compoundchem.com/2014/06/01/newoldbooksmell

"Why Do Books Smell So Good", *Science ABC*, 16 July 2015, https://www.scienceabc.com/nature/why-do-books-smell-so-good.html

15 EMERGENCY EMERGENCY PHONE CALLS

"About WPS", US Department of Homeland Security, 2 June 2017, https://www.dhs.gov/about-wps

"Making a Wireless Priority Service (WPS) Call", US Department of Homeland Security Office of Emergency Communications, May 2017, https://www.dhs.gov/sites/default/files/publications/Making%20a%20Wireless%20Priority%20Service%20%28WPS%29%20Call%20_May%202017%20FINAL%20508C.pdf

"Wireless Priority Service", US Department of Homeland Security Office of Emergency Communications, March 2017, https://www.dhs.gov/sites/default/files/publications/Wireless%20Priority%20Service_March%202017_FINAL%20508C%20031617%20%28003%29.pdf

"Wireless Priority Service (WPS) Eligibility Criteria", US Department of Homeland Security Office of Emergency Communications, May 2013, https://www.dhs.gov/sites/default/files/publications/FINAL%20WPS_Eligibility%20050513.pdf

"WPS Costs", US Department of Homeland Security, 20 October 2015, https://www.dhs.gov/wps-costs

"WPS Eligibility", US Department of Homeland Security, 5 June 2017, https://www.dhs.gov/wps-eligibility

16 LUSCIOUS LIPS

"A Quantitative Approach to Determining the Ideal Female Lip Aesthetic and Its Effect on Facial Attractiveness", by N.A. Popenko et al., *Journal of the American Medical Association Facial Plastic Surgery*, 16 February 2017, DOI:10.1001/jamafacial.2016.2049

"Determining the 2-Dimensional Threshold for Perception of Artificial-Appearing Lips", by Sang W. Kim and Daniel E. Rousso, *Journal of the American Medical Association Facial Plastic Surgery*, 6 April 2017, DOI:10.1001/jamafacial.2017.0052

17 SOLAR ENERGY PAYBACK TIME

"Compared Assessment of Selected Environmental Indicators of Photovoltaic Electricity in OECD Cities", The International Energy Agency–Photovoltaic Power Systems Programme, May 2006.

"Solar Cells", by Nicola Phillips, *The Science Show*, 5 April 2008, http://www.abc.net.au/radionational/programs/scienceshow/solar-cells/3263262

"Re-Assessment of Net Energy Production and Greenhouse Gas Emissions Avoidance after 40 Years of Photovoltaics Development", by Atse Louwen et al., *Nature Communications*, 6 December 2016, DOI:10.1038/ncomms13728

18 POKIES, TURNING THE TABLES

"Berejiklian Government Accused of 'Starving' Fairfield of Pokie Funds", by Sean Nicholls, *The Sydney Morning Herald*, 12 July 2017, http://www.smh.com.au/nsw/berejiklian-government-accused-of-starving-fairfield-of-pokie-funds-20170711-gx93t4.html

"Call to Publish Pubs and Clubs Pokie Profits as Gambling Surges", by Sean Nicholls, *The Sydney Morning Herald*, 22 February 2017, http://www.

REFERENCES

smh.com.au/nsw/call-to-publish-pubs-and-clubs-pokie-profits-as-gambling-surges-20170221-guhvbd.html

"Episode 773: Slot Flaw Scofflaws", by Keith Romer and Nick Fountain, *NPR Planet Money*, 24 May 2017, http://www.npr.org/sections/money/2017/05/24/529865107/episode-773-slot-flaw-scofflaws

"Fairfield Council Calls For Poker Machine Freeze in High-risk Areas", by Sean Nicholls, *The Sydney Morning Herald*, 11 July 2017, http://www.smh.com.au/nsw/fairfield-council-calls-for-poker-machine-freeze-in-high-risk-areas-20170710-gx85el.html

"Gambling with Pokies: Melbourne Gamblers Plug $2.6 Billion into Poker Machines Amid Record Losses", by Alex White, *Herald Sun*, 28 July 2017, http://www.heraldsun.com.au/news/victoria/gambling-with-pokies-%E2%80%A6-amid-record-losses/news-story/37d4e9be5cfb-15b9096c28071ca9937c

"Hot Lotto Fraud Scandal", *Wikipedia*, 27 January 2017, https://en.wikipedia.org/wiki/Hot_Lotto_fraud_scandal

"I Lost $500,000 Living in Fairfield, NSW's Ground Zero for Predatory Gambling", by Ralph Bristow, *The Sydney Morning Herald*, 17 July 2017, http://www.smh.com.au/comment/i-lost-500000-living-in-fairfield-nsws-ground-zero-for-predatory-gambling-20170716-gxc3x7.html

"Poker Machine Profits Come from Sydney's Poorest Suburbs" by James Robertson, *The Sydney Morning Herald*, 26 October 2015, http://www.smh.com.au/nsw/poker-machine-profits-come-from-sydneys-poorest-suburbs-20151023-gkh7j4.html

"Pokie Gambling in NSW Surges to $73 Billion Despite Fewer Machines", by Sean Nicholls, *The Sydney Morning Herald*, 21 November 2015, http://www.smh.com.au/nsw/pokie-gambling-in-nsw-surges-to-73b-despite-fewer-machines-20151119-gl2q1b.html

"Russians Engineer a Brilliant Slot Machine Cheat – And Casinos Have No Fix", by Brendan Koerner, *Wired*, 6 February 2017, https://www.wired.com/2017/02/russians-engineer-brilliant-slot-machine-cheat-casinos-no-fix

"Time for NSW Pokie Barons to Reveal Their Profits", by Sean Nicholls, *The Sydney Morning Herald*, 3 March 2017, http://www.smh.com.au/comment/time-for-nsw-pokie-barons-to-reveal-their-profits-20170302-guot9h.html

"Why Gladys Berejiklian Should Listen, At Least, to Fairfield on Pokies", *The Sydney Morning Herald*, 13 July 2017, http://www.smh.com.au/comment/smh-editorial/why-gladys-berejiklian-should-listen-at-least-to-fairfield-on-pokies-20170712-gx9rlx.html

19 GOLD NOBEL MEDAL INVISIBLE TO NAZIS

Adventures in Radioisotope Research: The Collected Papers of George Hevesy in Two Volumes, by George Hevesy, Pergamon Press, Oxford, 1962, pages 27–28.

"Dissolve My Nobel Prize! Fast! (A True Story)", by Robert Krulwich, NPR, 3 October 2011, http://www.npr.org/sections/krulwich/2011/10/03/140815154/dissolve-my-nobel-prize-fast-a-true-story

The Disappearing Spoon and Other True Tales of Madness, Love and the History of the World from the Periodic Table of the Elements, by Sam Kean, Little, Brown, New York, 2010, pages 212–215.

20 SPACE JUNK

"A Concept For Elimination of Small Orbital Debris", by Gurudas Ganguli et al., *arXiv.org*, 24 January 2012, https://arxiv.org/abs/1104.1401

"Allied Awareness", by Michael A. Taverna and Robert Wall, *Aviation Week & Space Technology*, 21 December 2009, page 36.

"Boom in Cheap Satellites Could Lead to 'Catastrophic Collision', Warn Scientists", by Ben Mitchell, *Independent*, 17 April 2017, http://www.independent.co.uk/news/science/new-satellites-collisions-catastrophic-communication-space-junk-a7687696.html

"Broadband Expansion Could Trigger Dangerous Surge in Space Junk" by Damien Gayle, *The Guardian*, 18 April 2017, https://www.theguardian.

REFERENCES

com/science/2017/apr/18/broadband-expansion-space-junk-internet-debris

"Crash Course", by Frank Morring Jr, et al., *Aviation Week & Space Technology*, 16 February 2009, pages 20–21.

"Debris Police", by Graham Warwick, *Aviation Week & Space Technology*, 6–19 March 2017, page 23.

"For Space Mess, Scientists Seek Celestial Broom", by Kenneth Chang, *The New York Times*, 18 February 2012, http://www.nytimes.com/2012/02/19/science/space/for-space-mess-scientists-seek-celestial-broom.html

"Global Warming Prolongs Life of Space Junk", by Adam Tanner, *ABC Science*, 12 December 2006, http://www.abc.net.au/science/news/stories/2006/1809502.htm?space

"Greenhouse Effect Could Cause a Space Problem", by Marc Kaufman, *The Washington Post*, 12 December 2009, http://www.washingtonpost.com/wp-dyn/content/article/2006/12/11/AR2006121101241_pf.html

"ISS Crew Takes Shelter", *Aviation Week & Space Technology*, 16 March 2009, page 18.

"Lost in Space: Debris Shield Bag Floats Away from Astronauts during ISS Spacewalk", *The Guardian*, 31 March 2017, https://www.theguardian.com/science/2017/mar/31/lost-in-space-debris-shield-bag-floats-away-from-astronauts-during-iss-spacewalk

"NASA Forced to Steer Clear of Junk in Cluttered Space", by William J. Broad, *The New York Times*, 31 July 2007, http://www.nytimes.com/2007/07/31/science/space/31orbi.html

"Operation Burnt Frost Team Is 2009 Operations Laureate", Amy Butler, *Aviation Week & Space Technology*, 16 March 2009, page 54.

"Operations Laureate", *Aviation Week & Space Technology*, 30 March 2009, page 54.

"Orbital Missions Safety – A Survey of Kinetic Hazards", by Vitaly Adushkin et al., *Acta Astronautica*, September–October 2016, Vol. 126, pages 510–516.

"Orbiting Dust Storm Could Remove Space Junk", *Technology Review*,

12 April 2011, https://www.technologyreview.com/s/423629/orbiting-dust-storm-could-remove-space-junk

"Shuttle and Space Station Dodge Debris", by Robert Block, *Los Angeles Times*, 23 March 2009, http://articles.latimes.com/2009/mar/23/nation/na-space-station23

"Space Age Wasteland: Debris in Orbit Is Here to Stay", by John Matson, *Scientific American*, 1 April 2012, https://www.scientificamerican.com/article/space-age-wasteland

"Space Cleanup", by Richard Tuttle, *Aviation Week & Space Technology*, 30 March 2009, page 38.

"Space Junk", by Frank Morring Jr, *Aviation Week & Space Technology*, 30 March 2009, page 14.

"Space Junk Apocalypse: Just like Gravity?", by Corrinne Burns, *The Guardian*, 15 November 2013, https://www.theguardian.com/science/blog/2013/nov/15/space-junk-apocalypse-gravity

"Space Junk Is Becoming a Real Problem: If We Don't Fix It, We Could Be Trapped on Earth", by Jennifer Harrison, *Gadgette*, 27 January 2016, http://www.gadgette.com/2016/01/27/space-junk-is-becoming-a-real-problem

"The ISS Astronauts Had to Shelter from Russian Space Junk Overnight", by Kelsey Campbell-Dollaghan, *Gizmodo*, 17 July 2015, https://www.gizmodo.com.au/2015/07/the-iss-astronauts-had-to-shelter-from-russian-space-junk-overnight

"'We've Left Junk Everywhere': Why Space Pollution Could Be Humanity's Next Big Problem", by Melissa Davey, *The Guardian*, 26 March 2017, https://www.theguardian.com/science/2017/mar/26/weve-left-junk-everywhere-why-space-pollution-could-be-humanitys-next-big-problem

21 CHILDHOOD AMNESIA

"A Price to Pay for Adult Neurogenesis", by Lucas A. Mongiat and Alejandro F. Schinder, *Science*, 9 May 2014, Vol. 344, No. 6184, pages 594–595.

"Childhood Memories Erased by Growth of New Brain Cells", by Clare

REFERENCES

Wilson, *New Scientist*, 8 May 2014, https://www.newscientist.com/article/dn25540-childhood-memories-erased-by-growth-of-new-brain-cells

"Hippocampal Neurogenesis Regulates Forgetting During Adulthood and Infancy", by Katherine G. Akers et al., *Science*, 9 May 2014, Vol. 344, No. 6184, pages 598–602.

"New Brain Cells Could Be Blanking Out Old Memories, Research on Rodents Suggests", by Meeri Kim, *The Washington Post*, 17 May 2014, https://www.washingtonpost.com/national/health-science/new-brain-cells-could-be-blanking-out-old-memories-research-on-rodents-suggests/2014/05/17/e2623844-dc6d-11e3-b745-87d39690c5c0_story.html

"The Mystery of Why You Can't Remember Being a Baby", by Zaria Gorvett, *BBC Future*, 26 July 2016, http://www.bbc.com/future/story/20160726-the-mystery-of-why-you-cant-remember-being-a-baby

"Why Can't We Remember Our Early Childhood?", by Jeanne Shinskey, *The Conversation*, 13 July 2016, http://theconversation.com/why-cant-we-remember-our-early-childhood-62325

"Why Can't You Remember Being a Baby?", by Annie Sneed, *Scientific American*, 15 July 2014, https://www.scientificamerican.com/article/why-can-t-you-remember-being-a-baby/

"Why Can't You Remember Being a Baby?", by Julia Davis, *Mental Floss*, 22 August 2012, http://mentalfloss.com/article/12330/why-cant-you-remember-being-baby

22 NATURAL ALARM CLOCK

"Timing the End of Nocturnal Sleep", by Jan Born et al., *Nature*, 7 January 1997, Vol. 397, No. 6714, pages 29–30.

23 SINKHOLES

"Dan Doctor on the May 2010 Sinkhole in Guatemala City", by Lindsay Patterson, *EarthSky*, 7 July 2010, http://earthsky.org/earth/dan-doctor-on-the-guatemala-city-sinkhole-of-may-2010

"Fire and Rescue Evacuate Residents in Sinkhole Drama in Illawong, Southern Sydney", by Leigh van den Broeke, *The Daily Telegraph*, 15 June 2015, http://www.dailytelegraph.com.au/news/news-story/ce59c834a404aac62ee1dedeefbc0d0f

"Sinkholes in Australia: Where and When Can They Strike?", by Alex Cullen, 27 April 2014, https://au.news.yahoo.com/sunday-night/features/a/22907841/sinkholes-in-australia-where-and-when-can-they-strike

"The Science of Sinkholes" by Jessica Robertson and Randall Orndorff, *US Geological Survey Science* Features blog, 11 March 2013, https://www2.usgs.gov/blogs/features/usgs_top_story/the-science-of-sinkholes

"What Are Sinkholes, and What Causes Them?", by Jon Henley, *The Guardian*, 5 March 2013, https://www.theguardian.com/world/2013/mar/04/what-causes-sinkholes-florida-man

24 EARTH STOPS SPINNING

"If the Earth Stood Still: Modeling the Absence of Centrifugal Force", by Witold Fraczek, *ArcUser*, Summer 2010, http://www.esri.com/news/arcuser/0610/nospin.html

"Your Science Questions Answered", by Nicola Davis, *The Observer*, 12 January 2014, https://www.theguardian.com/technology/2014/jan/12/your-science-questions-answered

"Early Earth Was Covered in Global Ocean and Had No Mountains", by New Scientist staff and Press Association, *New Scientist*, 8 May 2017, https://www.newscientist.com/article/2130266-early-earth-was-covered-in-a-global-ocean-and-had-no-mountains/

25 CANE TOADS CONFIRM CONCEPTION

"Rabbits, Mice and Toads: Testing for Pregnancy", by Caroline de Costa and Tony Kloss, *O&G Magazine*, Winter 2017, Vol. 19, No. 2, pages 68–70.

26 CHRONIC LATENESS

"Can You Cure Chronic Lateness?", by Li Zhou, *The Atlantic*, 2 June 2016, https://www.theatlantic.com/health/archive/2016/06/can-you-cure-lateness/485108

"Different Personalities Experience Time Differently", by Orion Jones, *Big Think*, 13 May 2015, http://bigthink.com/ideafeed/different-personalities-experience-time-differently

"How to Stop Being Late to Everything", by Sarah Elizabeth Richards, *Elle*, 18 July 2014, http://www.elle.com/culture/career-politics/how-to/a14690/four-steps-to-punctuality

"Incremental Validity of Time Urgency and Other Type A Subcomponents in Predicting Behavioural and Health Criteria", by Jeffrey M. Conte et al., *Journal of Applied Social Psychology*, August 2001, Vol. 31, No. 8, pages 1727–1748.

"Scientists Have Found Out Why You're Chronically Late" by Fiona MacDonald, *ScienceAlert*, 6 February 2015, http://www.sciencealert.com/scientists-have-identified-the-personality-type-that-makes-you-chronically-late

"Why I'm Always Late", by Tim Urban, *Wait But Why*, 7 July 2016, https://waitbutwhy.com/2015/07/why-im-always-late.html

"Why You're Always Late", by Sarah Elizabeth Richards, *Elle*, 6 August 2014, http://www.elle.com/life-love/advice/a14715/how-not-to-be-late

27 PB/5 PEDESTRIAN BUTTON

"Audio-Tactile Pedestrian Push Button Signalling System", Amalgamated Wireless Limited, US Patent 4851836 A, 25 July 1989.

"Sublime Design: The PB/5 Pedestrian Button", by Miles Park, *The Conversation*, 30 June 2014, https://theconversation.com/sublime-design-the-pb-5-pedestrian-button-26232

28 BEER "BEATS" CANCER?

"Beer Marinade Cuts Grilling Carcinogens", by Christopher Intagliata,

Scientific American, 4 April 2014, https://www.scientificamerican.com/podcast/episode/meat-beer-marinade/?sf82395952=1

"Dietary Benzo[a]Pyrene Intake and Risk of Colorectal Adenoma", by Rashmi Sinha et al., *Cancer Epidemiology, Biomarkers & Prevention*, August 2005, Vol. 14, No. 8, pages 2030–2034.

"Inhibitory Effect of Antioxidant-Rich Marinades on the Formation of Heterocyclic Aromatic Amines in Pan-Fried Beef", by Olga Viegas et al., *Journal of Agricultural and Food Chemistry*, 20 June 2012, Vol. 60, No. 24, pages 6235–6240.

"Why You Should Eat a Plant-Based Diet, But That Doesn't Mean Being a Vegetarian", by Katherine Livingstone, *The Conversation*, 13 July 2017, https://theconversation.com/why-you-should-eat-a-plant-based-diet-but-that-doesnt-mean-being-a-vegetarian-78470

29 MARIJUANA FOR MEMORY & LEARNING?

"A Chronic Low Dose of Delta-9-Tetrahydrocannabinol (THC) Restores Cognitive Function in Old Mice", by Andras Bilkei-Gorzo et. al., *Nature Medicine*, June 2017, Vol. 23, No. 6, pages 782–787.

"A Little Cannabis Every Day Might Keep Brain Ageing at Bay", by Michael Le Page, *New Scientist*, 8 May 2017, https://www.newscientist.com/article/2130257-a-little-cannabis-every-day-might-keep-brain-ageing-at-bay

"Daily Dose of Cannabis Extract Could Reverse Brain's Decline in Old Age, Study Suggests", by Ian Sample, *The Guardian*, 9 May 2017, https://www.theguardian.com/science/2017/may/08/daily-dose-of-cannabis-extract-could-reverse-brains-decline-in-old-age-study-suggests-thc

"Marijuana Improves Memory in Old Mice", nature.com Research Highlights, 8 May 2017, https://www.nature.com/articles/n-12271104

"Marijuana May Boost, Rather Than Dull, the Elderly Brain", by Stephani Sutherland, *Scientific American*, 10 May 2017, https://www.scientificamerican.com/article/marijuana-may-boost-rather-than-dull-the-elderly-brain/

REFERENCES

"The Cannabis Paradox: When Age Matters", by Andrés Ozaita and Ester Aso, *Nature Medicine*, June 2017, Vol. 23, No. 6, pages 661–662.

30 HOTEL MAGNETIC CARD ERASURE

"Explaining the Physics Behind Magstripes/Experimental Conclusions", by Jenny Magnes, *Lasers, Technology and Teleportation with Prof. Magnes*, 27 April 2011, http://pages.vassar.edu/ltt/?p=965 "The Magnetic Stripe Technology", by Shame Er Shah Kamal, *Illumin*, Fall 2006, Vol. VIII, No. II, http://illumin.usc.edu/157/the-magnetic-stripe-technology

"Historical Overview of the Card Industry", by Steven G. Halliday, *High Tech Aid*, 27 April 1998, http://www.hightechaid.com/tech/card/card_history.htm

"Introduction to Magnetic Stripe & Other Card Technologies", by Steven G. Halliday, *High Tech Aid*, 24 April 1997, http://www.hightechaid.com/tech/card/intro_ms.htm

"Magnetic Strips: How Do Magnetic Strips on Credit Cards Work?", by Ashish, *Science ABC*, 2 July 2017, https://www.scienceabc.com/innovation/how-magnetic-strips-stripes-tape-on-credit-room-key-cards-work.html

"What's With This Magnetic Stripe Stuff?", by Steven G. Halliday, *High Tech Aid*, http://www.hightechaid.com/tech/card/what_ms.htm

31 IG NOBEL PRIZES 2016

"An Unexpected Advantage of Whiteness in Horses: The Most Horsefly-Proof Horse Has a Deep Polarising White Coat", by Gábor Horváth et al., *Proceedings of the Royal Society B*, 3 February 2010, DOI:10.1098/rspb.2009.2202

"Being a Beast by Charles Foster Review – The Man Who Ate Worms Like a Badger", by Patrick Barkham, *The Guardian*, 3 February 2016, https://www.theguardian.com/books/2016/feb/03/being-beast-charles-foster-review-man-whoate-worms-like-badger

"Contraceptive Efficacy of Polyester-Induced Azoospermia in Normal

Men", by Ahmed Shafik, *Contraception*, May 1992, Vol. 45, No. 5, pages 439–451.

"Ecological Traps for Dragonflies in a Cemetery: The Attraction of Sympetrum Species (Odonata: Libellulidae) by Horizontally Polarising Black Gravestones", Gábor Horváth et al., *Freshwater Biology*, 27 June 2007, Vol. 52, No. 9, pages 1700–1709.

"Effect of Different Types of Textiles on Sexual Activity. Experimental Study", by Ahmed Shafik, *European Urology*, 1 January 1993, Vol. 24, No. 3, pages 375–380.

"From Junior to Senior Pinocchio: A Cross-Sectional Lifespan Investigation of Deception", by Evelyne Debye et al., *Acta Psychologica*, September 2015, Vol. 160, pages 58–68.

"Ig Nobel Awards Given for Rat Pants, Fly Catchers and Rocks with Personalities", by Camila Domonoske, *NPR*, 23 September 2016, http://www.npr.org/sections/thetwo-way/2016/09/23/495164220/ig-nobel-awards-given-for-rat-pants-fly-catchers-and-rocks-with-personalities

"Ig Nobel Honours Go to a Testicle Sling Contraceptive, Mirrored Itching", by John Timmer, *Ars Technica*, 24 September 2016, https://arstechnica.com/science/2016/09/testicle-sling-artificial-goat-legs-and-branding-of-rocks-honored/

"Ig Nobel Prizes: Trousers for Rats and the Truthfulness of Liars", by Alan Yuhas, *The Guardian*, 23 September 2016, https://www.theguardian.com/science/2016/sep/22/ig-nobel-prizes-trousers-for-rats-and

"In 'Bonk', Mary Roach Explores Science of Sex", *NPR*, 9 April 2008, http://www.npr.org/templates/story/story.php?storyId=89498532

"Itch Relief by Mirror Scratching. A Psychophysical Study", by Christoph Helmchen et al., *PLoS*, December 2013, Vol. 8, No. 12, e82756.

"Meet the Winners of This Year's Ig Nobel Prizes", by Jennifer Ouellette, 24 September 2016, https://www.gizmodo.com.au/2016/09/meet-the-winners-of-this-years-ig-nobel-prizes

"On the Reception and Detection of Pseudo-Profound Bullshit", by Gordon

REFERENCES

Pennycook et al., *Judgment and Decision-Making*, November 2015, Vol. 10, No. 6, pages 549–563.

"Perceived Size and Perceived Distance of Targets Viewed from Between the Legs: Evidence for Proprioceptive Theory", by Atsuki Higashiyama and Kohei Adachi, *Vision Research*, November 2006, Vol. 46, No. 23, pages 3961–3976.

"The 2016 Ig Nobel Prizewinners", Improbable Research, 24 September 2016, http://www.improbable.com/ig/winners/#ig2016

"The Brand Personality of Rocks: A Critical Evaluation of a Brand Personality Scale", by Mark Avis et al., *Marketing Theory*, 6 December 2013, DOI: 10.1177/1470593113512323

"'The Fly Trap,' by Fredrik Sjoberg", by Hugh Raffles, *The New York Times*, 31 July 2015, https://www.nytimes.com/2015/08/02/books/review/the-fly-trap-by-fredrik-sjoberg.html

"The Uppermost Aristocracy of the Hoverfly Society", by Byrd Pinkerton, *NPR*, 23 July 2016, http://www.npr.org/2016/07/23/486731588/the-uppermost-aristocracy-of-the-hoverfly-society

"US VW Probe Finds Criminal Wrongdoing, Regulators Work to Settle", Megan Geuss, *Ars Technica*, 16 August 2016, https://arstechnica.com/cars/2016/08/us-volkswagen-probe-finds-criminal-wrongdoing-regulators-working-to-settle

"Volkswagen Engineer Pleads Guilty to Conspiracy in Emissions Scandal", by Jana Kasperkevic, *The Guardian*, 10 September 2016, https://www.theguardian.com/business/2016/sep/09/volkswagen-engineer-pleads-guilty-conspiracy-emissions-scandal-

"What Does the Goat Man Say? Baa, Maa, or 'I'm Crazy'", by Marc Silver, *NPR* Goats and Soda blog, 22 May 2016, http://www.npr.org/sections/goatsandsoda/2016/05/22/478719168/what-does-the-goat-man-say-baa-maa-or-im-crazy

"What Makes People Susceptible to Pseudo-Profound 'Baloney'?", by Tania Lombrozo, *NPR* 13.7 Cosmos & Culture, 7 December 2015, http://www.

npr.org/sections/13.7/2015/12/07/458740250/what-makes-people-susceptible-to-pseudo-profound-baloney

32 WIFI SPY

"3D Tracking via Body Radio Reflections", by Fadel Adib et al., *Proceedings of the 11th USENIX Conference on Networked Systems Design and Implementation*, 2–4 April 2014, pages 317–329.

"All the Ways Your WiFi Router Can Spy on You", by Kaveh Waddell, *The Atlantic*, 24 August 2016, https://www.theatlantic.com/technology/archive/2016/08/wi-fi-surveillance/497132

"Capturing the Human Figure through a Wall", by Fadel Adib et al., *ACM Transactions on Graphics (TOG) – Proceedings of ACM SIGGRAPH Asia 2015*, November 2015, Vol. 34, No. 6, article no. 219.

"Demo: Real-Time Breath Monitoring Using Wireless Signals", by Fadel Adib et al., *Proceedings of the 20th Annual International Conference on Mobile Computing and Networking*, 7–11 September 2014, pages 261–262.

"FreeSense: Indoor Human Identification with WiFi Signals", by Tong Xin et al., *arXiv.org*, 11 August 2016, https://arxiv.org/ftp/arxiv/papers/1608/1608.03430.pdf

"Keystroke Recognition Using WiFi Signals", by Kamran Ali et al., *Proceedings of the 21st Annual International Conference on Mobile Computing and Networking*, 7–11 September 2015, pages 90–102.

"Multi-Person Localisation via RF Body Reflections", by Fadel Adib et al., *Proceedings of the 12th USENIX Symposium on Networked Systems Design and Implementation*, 4–6 May 2015, pages 279–292.

"See Through Walls with Wi-Fi!", by Fadel Adib and Dina Katabi, *Proceedings of the ACM SIGCOMM 2013 Conference on SIGCOMM*, 12–16 August 2013, Hong Kong, pages 75–86.

"Smart Homes that Monitor Breathing and Heart Rate", by Fadel Adib et al., *Proceedings of the 33rd Annual ACM Conference on Human Factors in Computing Systems*, 18–23 April 2015, pages 837–846.

"We Can Hear You with Wi-Fi!", by Guanhua Wang et al., *Proceedings of the 20th Annual International Conference on Mobile Computing and Networking*, 7–11 September 2014, pages 593–604.

"What Does the "Fi" in Wi-Fi Mean?", by Akemi Iwaya, *How-To Geek*, 14 June 2016, https://www.howtogeek.com/259000/what-does-the-fi-in-wi-fi-mean

"WiFi-ID: Human Identification Using WiFi Signal", by Jin Zhang et al., *2016 Proceedings of the International Conference on Distributed Computing in Sensor Systems (DCOSS)*, 26–28 May 2016. pages 75–82.

33 SPIDER WEBS DON'T TWIST

"Biopolymers: Shape Memory in Spider Draglines", by Olivier Emile et al., *Nature*, 30 March 2006, Vol. 440, No. 7084, page 621.

"Peculiar Torsion Dynamical Response of Spider Dragline Silk", by Dabiao Liu, et al.., *Applied Physics Letters*, July 2017, Vol. 111, No. 1, DOI: http://dx.doi.org/10.1063/1.4990676

"Strange Silk: Why Rappelling Spiders Don't Spin out of Control," Phys.org, 7 July 2017, https://phys.org/news/2017-07-strange-silk-rappelling-spiders-dont.html

"Research Find Why Rappelling Spiders Spin in Control", by Amy Wallace, UPI.com, 7 July 2017, http://www.upi.com/Science_News/2017/07/07/Research-finds-why-rappelling-spiders-spin-in-control/3941499445551

"Why Abseiling Spiders Don't Spin out of Control", by Sarah Tesh, physicsworld.com, 13 July 2017, http://physicsworld.com/cws/article/news/2017/jul/13/why-abseiling-spiders-dont-spin-out-of-control

"Why Spiders' Silk Threads Don't Twist", Phys.org, 30 March 2006, https://phys.org/news/2006-03-spiders-silk-threads-dont.html

THANK YOUS

Books take time to write. Like Space Travel, book writing is a Long Game. I'm lucky to have received much kindness and help along the way – and not from strangers.

Mary loves whipping out the red pen and, crazy with the power, she shapes my words till they actually express what I am trying to say.

Then come the punchlines, which are definitely a labour of love, so Big Thanks to Mary, Isabelle Benton, and my ABC Producers, Daniel Driscoll, Tiger Webb and Bernie Hobbs, who all helped make the stories more approachable and less academic. My Yadults (youth grown up into adults), Alice, Lola and non-Little Karl, are full of valuable suggestions and sometimes blunt insights, and I love them and their feedback.

This is my first year working with Jules Faber as illustrator. He is wonderfully simpatico. Jules has given this lovely book both whimsical/comedic and "technical" illustrations – a tricky balance, but he has excelled.

Thanks also to everyone who proofreads for me. Every year, even at the very last stages, we always find mistakes. Special thanks here go to my Fabulous Editor, Danielle Walker; my Fabulous Producer, Isabelle Benton; my Fabulous Agent, Jo Butler; and my Fabulous Noise Boy at the ABC, Chris Stedman – who went above and beyond the Call of Duty and spent days beforehand going through every letter/number/punctuation mark/fact/claim (and fact!).

My special power is that I am a moderately broad Generalist, but I am not expert in any field whatsoever – so Big Thanks to the invaluable Experts. (As always – all mistakes are my fault; all goodness is from them.) So, Big It Up for Alice Williamson ("Brave New Womb"), Clare Collins (Coffee again), Paul Schofield ("Emergency Emergency Phone Calls"), Marc Abrahams ("Ig Nobel Prizes 2016"), Jessica Bloom (a Real Physicist, who gave me the low-down on Ponytail Dynamics), Brendan Gregg ("Shouting at Hard Drives"), Jason Held ("Space Junk"), Geraint Lewis (Gravitational Waves), and Jin Zhang and Salil S. Kanhere ("WiFi Spy").

Words travel a long path from a computer screen until finally turning into a book. So, on the Publishing side, let me thank Claire Craig (most patient publisher), Jo Butler (super-agent), Danielle Walker and Sarah Fletcher (epic editors), and Alissa Dinallo (designer).

Finally, let me thank my ever so tolerant family for letting me play my Writing Music, "Imitation of the Bells" ("*Imitazione delle campane*") over 9,000 times so my little book could take off from cyberspace to a form more affected by gravity . . .

Also by
DR KARL KRUSZELNICKI

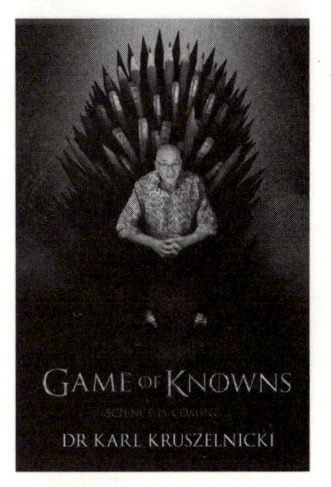

Also by
DR KARL KRUSZELNICKI

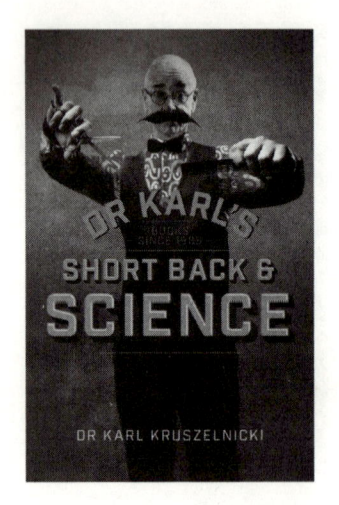

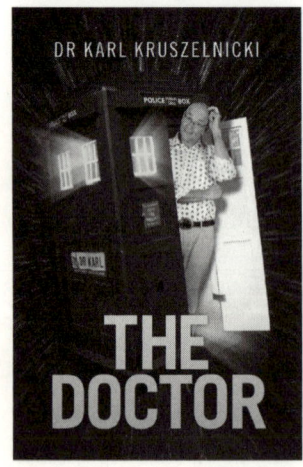